Agriculture in the United Kingdom: 1991

LONDON: HMSO

© Crown copyright 1992
Applications for reproduction should be made to HMSO
First published 1992

ISBN 0 11 242919 X

Contents

Preface	*page*	vii
1	**Summary of the year**	1
2	**The structure of the industry**	5
3	**Policy developments in 1991**	15
4	**Output prices and input costs**	18
5	**Commodities**	20
6	**Agricultural incomes**	49
7	**Land prices and balance sheets**	62
8	**Farm business data**	64
9	**Public expenditure on agriculture**	70

Statistical tables and charts

List of tables

1.1	Agriculture and food in the national economy	*page* 4
2.1	Agricultural land use	7
2.2	Crop areas and livestock numbers	8
2.3	Numbers and sizes of holdings	10
2.4	Numbers and sizes of enterprises	11
2.5	Labour force in agriculture	14
2.6	Fixed capital stock of agriculture	14
4.1	Price indices for products and inputs	18
5.1	Wheat	22
5.2	Barley	23
5.3	Oats	24
5.4	Rye, mixed corn and triticale	24
5.5	Maize	25
5.6	Total cereals	25
5.7	Oilseed rape	28
5.8	Sugar beet and sugar	28
5.9	Hops	29
5.10	Peas and beans for stockfeed	29
5.11	Purchased feedingstuffs	30
5.12	Herbage and legume seeds	31
5.13	Purchased seeds	31
5.14	Potatoes	32
5.15	Horticulture	34
5.16	Selected horticultural crops	35
5.17	Cattle and calves; beef and veal	39
5.18	Sheep and lambs; mutton and lamb	40
5.19	Pigs and pigmeat	41
5.20	Poultry and poultrymeat	42
5.21	Milk	44
5.22	Milk products	45
5.23	Eggs	47
5.24	Wool	48
6.1	Outputs, inputs and net product	52
6.2	Changes in outputs and inputs	55
6.3	Output volume and productivity	57
6.4	Summary measures from the aggregate agricultural account	57

6.5	Gross capital formation	*page* 59
6.6	Stocks and work in progress	60
6.7	Costs and earnings of hired labour	60
6.8	Interest	61
6.9	Farm rents	61
7.1	Agricultural land prices	63
7.2	Aggregate balance sheets for agriculture	63
8.1	Net farm income by country and farm type	66
8.2	Net farm income by farm type, country and size	67
8.3	Occupier's net income by farm type, country and tenure	68
8.4	Assets and liabilities of farm businesses by country and tenure	69
9.1	Public expenditure under the CAP and on national grants and subsidies	72
9.2	Public expenditure under the CAP by the Intervention Board and the Agricultural Departments - major commodities	74
9.3	Commodity intervention in the UK	75

List of charts

2.1	Agricultural land use: 1991	5
2.2	Changes in crop areas and livestock numbers	9
4.1	Price indices for products and inputs	19
6.1	Outputs, inputs, net product and the resulting incomes	54
6.2	Changes in outputs and inputs	56
6.3	Trends in incomes from farming	58
9.1	Public expenditure under the CAP by the Intervention Board and the Agricultural Departments	70

Preface

1. This issue of Agriculture in the United Kingdom sets out data on the economic condition of, and prospects for, the United Kingdom agricultural industry considered during the 1991 annual review. The Government will draw on this information when considering policy issues, including proposals by the European Commission for reform of the Common Agricultural Policy and for agricultural support in 1992/93.

Statistical tables - general note

2. The tables in this issue are similar to those in the previous issue, Agriculture in the United Kingdom 1990. However some of the figures for past years may differ from those published in the preceding issues. This is because of the use of later information, changes in the scope and nature of the available data and improvements in statistical methods. A guide to the content and structure of the commodity tables is provided in the introduction to Section 5.

3. Most of the data are on a calendar year basis. The figures for 1991 are described as forecasts since they generally reflect the position up to the end of the year as seen at November 1991 on the basis of the information then available. The figures in the tables in Sections 8 and 9 relate to years ending (on average) in mid-February and at the end of March, respectively.

4. The following points apply throughout:
(i) all figures relate to the United Kingdom, unless otherwise stated;
(ii) the figures for imports and exports include those from intervention stocks and the figures for exports include re-exports. The figures for trade with the eleven other member states of the European Community and with third countries reflect country of consignment for imports and country of reported final destination for exports. The source of Overseas Trade Statistics is HM Customs and Excise;
(iii) where the units are common or predominant, they are indicated at the top of the table. Otherwise they are shown in the body of the table;
(iv) in some cases figures may not add to the corresponding totals because of roundings;
(v) symbols:
- means 'nil'
... means 'negligible' (less than half the last digit shown)
.. means 'not available' or 'not applicable'.

1 Summary of the year

Summary of the year

1. In the European Community, the year was dominated by discussions on the reform of the CAP; in the autumn the Commission made proposals for far reaching changes to a number of commodity regimes. Discussion of these is continuing. In the annual price negotiations most common prices were unchanged, with reductions for some commodities; the green pound was devalued to bring it into line with the market exchange rate. In October a new regime for oilseeds was agreed in principle.

2. In the UK a pilot quality wine scheme was launched in June. In July a Code of Good Agricultural Practice for the Protection of Water was launched, and new regulations on straw burning came into force. In November proposals for additional Environmentally Sensitive Areas were announced.

3. In November the Government published a policy statement, 'Our Farming Future'. This outlines the Government's thinking on the challenges and opportunities facing the industry in the 1990s.

4. The weather conditions during the 1990/91 agricultural year were generally favourable. The autumn of 1990 was mild in England but wetter than average in Scotland and Northern Ireland. For the first time in four years, the winter was not unusually mild and, with a severe cold spell in early February, this helped to discourage pests. By contrast, March was unusually warm but both May and June were cool and dull before temperatures rose to above normal levels in July and August. Following some fairly wet months (May apart), August and much of September were unusually dry, providing favourable conditions for harvesting.

5. The forecasts of the output, productivity and income of UK agriculture in 1991 show the following changes relative to 1990:

- cereal yields rose considerably, with those of wheat and barley being exceeded only by the levels attained in the exceptional year of 1984. However with a further decline in the planted area, total cereal production was little changed from the levels of the previous years. The value of cereal output rose by 6 per cent, largely as a result of higher wheat prices;

- more land was sown to oilseed rape but, with lower yields and prices, the value of output declined by 5 per cent. That of sugar beet also declined slightly;

- yields of early potatoes fell considerably but the fall was more than offset by much higher prices. The value of output of all potatoes was virtually unchanged since the area, yield and price for the maincrop all declined slightly;

- the value of output of vegetables fell by 5 per cent, mainly due to lower producer prices. Higher prices for some fruits offset a decline in the volume of output of fruit;

- higher production of both beef and sheep meat was reflected in increases in output values of 3 and 8 per cent respectively;

- production of pig meat also rose but lower prices resulted in a 6 per cent fall in the value of output;

- increased production of both poultry and eggs was offset by lower prices, with the value of poultry output being virtually unchanged but that of eggs falling by 8 per cent;

- milk production, which is constrained by quotas, fell and was reflected in a 2 per cent fall in the value of output;

- overall, the volume of the industry's gross output rose by about 1 per cent while that of its gross input fell slightly. Taking account of changes in all its inputs (including also both labour and fixed capital) the industry recorded a further 3 per cent increase in productivity;

- however with an increase in overall input prices of nearly 4 per cent, and a decline of 1 per cent in average realised returns from output, the industry's net product fell by nearly 5 per cent (in money terms);

- with the decline in interest rates over the year, the industry's interest charges fell by 14 per cent. However, despite a further fall in the number of farm workers, the total cost of hired labour rose by 3 per cent;

- total income from farming, representing the total income from agriculture of farmers, partners, directors and their spouses and family workers, is thus forecast to have fallen by 6 per cent. Farming income, which covers only farmers and their spouses, is forecast to have fallen by 14 per cent;

- however much lower falls, of only 1 and 4 per cent respectively, are forecast for the corresponding cash flow measures;

- all these changes are relative to the estimates for 1990 shown in this publication. The forecast levels of income and cash flow in 1990 published last year have all been revised upwards, matched by more favourable changes between 1989 and 1990;

- there were wide variations in the fortunes of different types of farm in the different parts of the United Kingdom. In most of the country cereal and cropping farms fared relatively well in 1991, but in Scotland the high yields of the previous year have not been repeated and incomes are forecast to fall. Incomes on livestock farms in the hills and uplands are generally expected to have improved as a result of substantially increased subsidy payments, while cuts in milk quota are expected to reduce incomes on dairy farms. The intensive livestock sector has been adversely affected by the sharp reduction in output prices for pigs and poultry products.

Agriculture and food in the national economy *(Table 1.1)*

6. These developments are described in more detail in the following sections. Table 1.1 provides a very broad picture of agriculture and food in the national economy.

TABLE 1.1 Agriculture and food in the national economy

Calendar years

	Average of 1980-82	1987	1988	1989	1990	1991 (provisional)
Agriculture's contribution to Gross Domestic Product (a)						
at current prices (£ million)	4,577	5,565	5,369	6,261	6,527	6,324
at constant 1985 prices (£ million)	4,534	4,934	4,867	5,206	5,536	5,692
% of national GDP (current prices)	2.1	1.6	1.4	1.4	1.4	1.3
Workforce in agriculture						
('000 persons)	639	588	580	567	565	552
% of total workforce in employment	2.6	2.3	2.2	2.1	2.1	2.1
Gross fixed capital formation in agriculture						
at current prices (£ million)	1,086	968	1,125	1,123	1,182	..
at constant 1985 prices (£ million)	1,212	899	965	882	893	..
% of national GFCF (current prices)	2.6	1.3	1.3	1.1	1.1	..
						(Jan - Oct)
Imports of food, feed and drink						
(£ million)(b)	6,814	10,276	10,602	11,420	12,298	10,080
of which: food, feed and						
non alcoholic drinks	6,352	9,236	9,470	10,195	10,861	8,990
alcoholic drinks	462	1,040	1,132	1,225	1,437	1,089
Volume index (1985=100)	94.2	113.1	115.2	118.1	122.1	121.5
Unit value (price) index (1985=100)	75.0	98.1	98.3	104.3	108.5	107.9
% of total UK imports	13.0	10.9	10.0	9.4	9.8	10.2
						(Jan - Oct)
Exports of food, feed and drink						
(£ million)(b)	3,170	5,166	4,909	5,886	6,352	5,412
of which: food, feed and						
non alcoholic drinks	2,221	3,778	3,361	4,118	4,286	3,675
alcoholic drinks	950	1,388	1,548	1,768	2,065	1,737
Volume index (1985=100)	85.6	122.0	116.0	123.8	128.6	126.9
Unit value (price) index (1985=100)	80.3	101.0	102.0	109.5	118.5	123.2
% of total UK exports	6.2	6.5	6.0	6.3	6.1	6.3
UK self-sufficiency in food and feed as a % of:						
all food and feed	60.5	56.8	53.9	58.5	56.5	58.2
indigenous type food and feed	75.7	72.3	68.9	75.5	72.1	73.8
Consumers' expenditure on food and drink						
at current prices (£ million)	42,367	63,923	70,647	76,662	83,263	89,400
of which: household food	25,030	34,472	36,593	39,245	41,833	44,200
meals out (c)	6,300	12,000	15,300	17,600	19,700	21,400
alcoholic drinks	11,037	17,451	18,754	19,817	21,730	23,800
at constant 1985 prices (£ million)	53,441	58,982	61,928	63,601	63,839	64,000
% of total consumers' expenditure	27.3	24.1	23.6	23.5	23.8	24.2
of which: household food	16.1	13.0	12.2	12.0	12.0	12.0
meals out (c)	4.1	4.5	5.1	5.4	5.6	5.8
alcoholic drinks	7.1	6.6	6.3	6.1	6.2	6.4
						(Jan - Nov)
Retail price indices (1985=100)						
food	82.5	106.4	110.1	116.4	125.7	132.0
alcoholic drinks	73.5	108.8	114.3	120.8	132.4	148.5
all items	78.5	107.7	113.0	121.8	133.3	140.9

(a) Agriculture is here defined as in the national accounts, that is net of the landlord element and the produce of gardens and allotments.
(b) The coverage of this aggregate is now that of SITC divisions 01-09, 11, 22 and section 4.
(c) Does not cover meals that are included with accommodation.

2 The structure of the industry

Introduction

1. The tables in this section portray the size and structure of the UK agricultural industry in 1991 and earlier years. Together they provide information on land use and livestock numbers in UK agriculture, on the distribution of these between holdings, and on the industry's labour force and its stock of fixed capital.

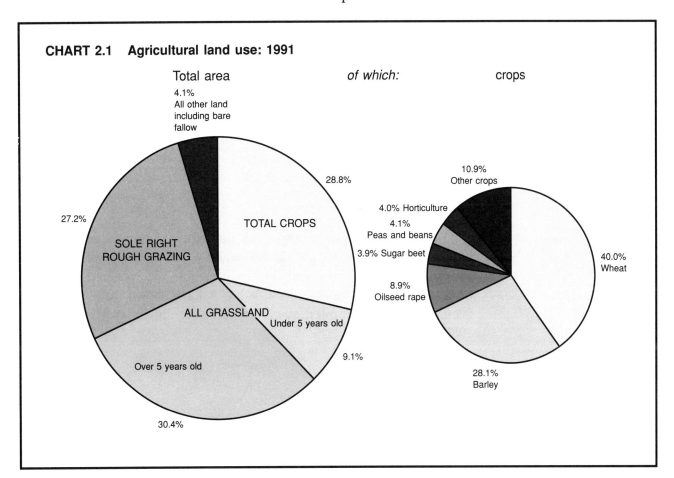

CHART 2.1 Agricultural land use: 1991

Crop areas and livestock numbers
(Tables 2.1 and 2.2)

2. At June 1991 the total area of agricultural land was 18.5 million hectares, some 77 per cent of the total land area in the UK. Details of the use of this land and of the main changes over the last decade are provided in Tables 2.1 and 2.2. Between 1990 and 1991, the area of arable land showed a small decline: there were decreases in the areas of all main cereals and in that of peas for harvesting dry and field beans, but the area of oilseed rape increased sharply for the second year in succession, as did linseed within the category of 'other crops'. The main trends in the livestock section of Table 2.2 are the continuing decline of the dairy herd, the marked upturn in the beef breeding herd over the last five years and a slowing down in the expansion of sheep

numbers. The pattern of agricultural land use in 1991 is shown in Chart 2.1 and the changes on 1990 in crop areas and livestock numbers in Chart 2.2.

Sizes of holdings and enterprises
(Tables 2.3 and 2.4)

3. Tables 2.3 and 2.4 give an insight into the structure of UK agriculture at June 1991. These tables exclude 'minor' holdings which are very small in terms of their area and activity, and hence relate solely to 'main' holdings. Table 2.3 shows that 30,000 holdings in the largest size group, just over 12 per cent of main holdings, accounted for 56.8 per cent of total agricultural activity. Size distributions of main holdings according to their crop areas and livestock numbers are presented in Table 2.4.

Labour *(Table 2.5)*

4. Table 2.5 records the number of persons engaged in UK agriculture at June of each year. The total labour force decreased by 2% between 1990 and 1991. The number of regular whole-time male workers has continued to decline, while the number of all other workers has been fairly stable since 1989. The overall number of farmers, partners and directors has continued to fall although, within this total, the number working part time has shown a further small increase.

Fixed capital stock
(Table 2.6)

5. Table 2.6 gives information on the stock of fixed capital (excluding land) available to the industry. The figures are at constant 1985 prices before allowing for depreciation. The estimates give an indication of the size of the industry's stock of productive assets and of how this has changed over recent years.

6. Agriculture's fixed capital stock, valued at 1985 prices, is estimated to have been £23,710 million at the end of 1990, just over 1 per cent down on the previous year. The reduction has been due to plant and machinery, while the stocks of buildings and vehicles have shown little change. The overall stock of fixed capital is estimated to be at a similar level to that in 1980-1982.

TABLE 2.1 Agricultural land use

The data in this table cover all holdings (including minor holdings) in England and Wales but exclude minor holdings in Scotland and Northern Ireland (a)

'000 hectares   At June of each year

	Average of 1980-82	1987	1988	1989	1990	1991 (provisional)
Total agricultural area (total area on agricultural holdings) plus common rough grazing	18,848	18,676	18,651	18,581	18,542	18,498
This comprises:						
Crops	5,013	5,271	5,255	5,137	5,013	4,974
Bare fallow	63	42	58	65	64	63
Total tillage	5,077	5,313	5,313	5,203	5,077	5,038
All grass under five years old	1,911	1,692	1,614	1,535	1,580	1,579
Total arable land	6,988	7,005	6,928	6,738	6,657	6,617
All grass five years old and over (excluding rough grazing)	5,113	5,110	5,159	5,249	5,263	5,248
Total crops and grass (b)	12,101	12,115	12,087	11,986	11,921	11,865
Sole right rough grazing	5,041	4,791	4,759	4,736	4,706	4,693
All other land on agricultural holdings including woodland (c)	491	554	570	623	680	707
Total area on agricultural holdings	17,634	17,460	17,415	17,345	17,307	17,265
Common rough grazing (estimated)	1,214	1,216	1,236	1,236	1,236	1,233

(a) Some of the figures shown in this table differ slightly from those in earlier editions of this publication since, for years from 1987 onwards, they now exclude returns on about 2,500 holdings (net) in Scotland which have been reclassified as minor holdings.
(b) Includes bare fallow.
(c) In Great Britain other land comprises farm roads, yards, buildings (excluding glasshouses), ponds and derelict land. In Northern Ireland other land includes land under bog, water, roads, buildings etc and wasteland not used for agriculture.

TABLE 2.2 Crop areas and livestock numbers

The data in this table cover all holdings (including minor holdings) in England and Wales but exclude minor holdings in Scotland and Northern Ireland (a)(b)

At June of each year

	Average of 1980-82	1987	1988	1989	1990	1991 (provisional)
Crop areas ('000 hectares)						
Total	5,013	5,271	5,255	5,137	5,013	4,974
This comprises:						
Total cereals	3,982	3,937	3,898	3,873	3,657	3,519
of which: wheat	1,532	1,994	1,886	2,083	2,013	1,990
barley	2,293	1,831	1,879	1,652	1,516	1,400
oats	140	98	120	118	107	105
rye and mixed corn	18	13	12	12	12	13
triticale (c)	..	..	..	8	9	11
Other arable crops (excluding potatoes)	604	957	968	882	971	1,078
of which: oilseed rape	130	388	347	321	390	445
sugar beet not for stockfeeding	209	202	201	197	194	196
hops	6	4	4	4	4	4
peas for harvesting						
dry and field beans	74	208	260	215	216	202
other crops	185	155	155	146	167	231
Potatoes	196	178	180	174	177	177
Horticulture	230	200	209	208	208	200
of which: vegetables grown in the open	153	132	141	141	142	136
orchard fruit	44	38	37	36	34	34
soft fruit	18	15	15	15	15	15
ornamentals (d)	13	12	13	14	14	14
other	2	2	2	2	2	2
Livestock numbers ('000 head)						
Total cattle and calves	13,269	12,170	11,884	11,975	12,059	11,871
of which: dairy cows	3,223	3,042	2,912	2,865	2,847	2,765
beef cows	1,429	1,345	1,375	1,495	1,599	1,670
heifers in calf	851	775	834	793	757	734
Total sheep and lambs	32,204	38,756	41,007	42,988	43,799	43,919
of which: ewes and shearlings	15,324	18,123	19,077	20,039	20,411	20,389
lambs under one year old	15,638	19,381	20,596	21,564	22,023	22,171
Total pigs	7,889	7,943	7,982	7,509	7,449	7,659
of which: sows in pig and other sows for breeding	730	713	703	660	660	676
gilts in pig	114	107	101	97	109	105
Total fowls (e)	125,264	128,801	130,998	120,351	124,615	126,365
of which: table fowls including broilers	59,274	70,869	75,437	70,176	73,588	75,228
laying fowls	45,092	38,529	37,420	33,957	33,468	33,209
growing pullets	14,481	12,238	11,243	9,414	10,452	10,586

(a) For this and other reasons, the crop area figures shown in this table may differ slightly from those shown in Section 5 which, in principle, cover all holdings as they are directly linked to the valuation of output.
(b) Some of the figures shown in this table differ slightly from those in earlier editions of this publication since, for years from 1987 onwards, they now exclude returns on about 2,500 holdings (net) in Scotland which have been reclassified as minor holdings.
(c) Collected separately for the first time in 1989 (Great Britain only).
(d) Hardy nursery stock, bulbs and flowers.
(e) Because of changes in the coverage of poultry holdings, data from 1987 cannot be directly compared with the average for 1980-1982.

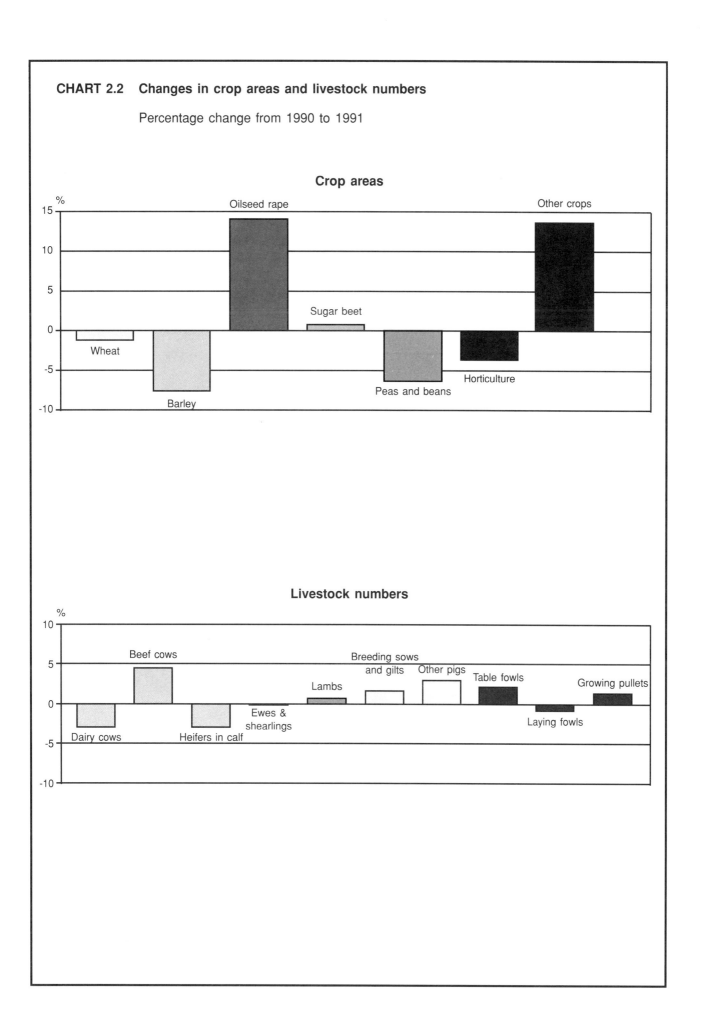

TABLE 2.3 Numbers and sizes of holdings

The data in this table exclude minor holdings

At June of each year

		1986		1991 (provisional) (a)	
		Number of holdings ('000)	Percent of total BSU	Number of holdings ('000)	Percent of total BSU
Size of holding (BSU) (b) (British Size Units (BSUs) measure the financial potential of the holding in terms of the margins which might be expected from its crops and stock. The margins used are gross margins standardised at average 1978-80 values. The threshold of 4 BSU is judged to be the minimum for full-time holdings).	under 4.0 BSU	102.8	2.7	102.6	2.6
	4.0 to under 16.0 BSU	65.6	13.5	60.9	12.5
	16.0 to under 40.0 BSU	50.4	29.0	48.0	28.0
	40.0 BSU and over	29.3	54.8	30.0	56.8
	Total	248.1	100.0	241.4	100.0
	Holdings 4 BSU and over:				
	Average size :BSUs	..	29.6	..	30.7
	:total area (hectares)	..	105.8	..	107.3

		Number of holdings ('000)	Hectares ('000)	Number of holdings ('000)	Hectares ('000)
Total area on holdings (b)	0.1 to under 20 hectares	102.2	834	97.9	820
	20.0 to under 50 hectares	62.8	2,073	60.4	2,001
	50.0 to under 100 hectares	43.1	3,051	42.7	3,032
	100.0 hectares and over	40.0	11,337	40.4	11,268
	Total	248.1	17,295	241.4	17,122
	Average total area per holding (hectares)	..	69.7	..	70.9
	% of total area on holdings with 100 hectares and over	..	65.6	..	65.8

		Number of holdings ('000)	Hectares ('000)	Number of holdings ('000)	Hectares ('000)
Crops and grass area (b)(c)	0.1 to under 20 hectares	102.0	827	96.2	803
	20.0 to under 50 hectares	63.0	2,079	60.3	1,991
	50.0 to under 100 hectares	41.2	2,904	40.9	2,877
	100.0 hectares and over	31.2	6,168	31.2	6,109
	Total	237.4	11,978	228.6	11,780
	Average crops and grass area per holding (hectares)	..	50.5	..	51.5
	% of total crops and grass area on holdings with 100 hectares and over	..	51.5	..	51.9

(a) The provisional 1991 figures include returns on about 3,500 holdings (net) in Scotland which have been reclassified as main holdings.
(b) Land in Great Britain let out under short term lets is attributed to the lessor, but land so let out in Northern Ireland (under the conacre system) is now attributed to the lessee. This difference, which applies to both the 1986 and 1991 figures in the table, affects both the number of holdings and their average size.
(c) The numbers of holdings shown in this part of the table are lower than those presented in the "total area" part of the table because holdings without crops and grass are excluded.

TABLE 2.4 Numbers and sizes of enterprises

The data in this table exclude minor holdings

At June of each year

		1986		1991 (provisional) (a)	
		Number of holdings ('000)	Hectares ('000)	Number of holdings ('000)	Hectares ('000)
Cereals (excluding maize)	0.1 to under 20 hectares	47.1	358	37.6	304
	20.0 to under 50 hectares	21.0	679	19.4	630
	50.0 hectares and over	24.1	2,982	21.7	2,572
	Total	92.2	4,019	78.7	3,506
	Average area (hectares)	..	43.6	..	44.5
	% of total cereals area on holdings with 50 hectares and over	..	74.2	..	73.4
Oilseed rape	0.1 to under 20 hectares	7.2	76	8.6	97
	20.0 to under 50 hectares	4.4	125	5.9	181
	50.0 hectares and over	1.4	98	2.0	160
	Total	13.0	299	16.5	438
	Average area (hectares)	..	23.0	..	26.5
	% of total oilseed rape area on holdings with 50 hectares and over	..	32.7	..	36.5
Sugar beet (England and Wales only)	0.1 to under 10 hectares	5.4	28	4.6	25
	10.0 to under 20 hectares	2.8	40	2.7	38
	20.0 hectares and over	3.1	137	3.0	133
	Total	11.4	206	10.3	195
	Average area (hectares)	..	18.0	..	18.9
	% of total sugar beet area on holdings with 20 hectares and over	..	66.8	..	67.9
Potatoes	0.1 to under 10 hectares	30.0	67	21.2	53
	10.0 to under 20 hectares	3.1	42	3.2	43
	20.0 hectares and over	1.9	68	2.2	81
	Total	35.0	177	26.6	177
	Average area (hectares)	..	5.1	..	6.6
	% of total potato area on holdings with 20 hectares and over	..	38.2	..	45.9

TABLE 2.4 Numbers and sizes of enterprises (continued)

The data in this table exclude minor holdings

At June of each year

		1986		1991 (provisional) (a)	
		Number of holdings ('000)	Number of livestock ('000)	Number of holdings ('000)	Number of livestock ('000)
Dairy cows	1 to 49 dairy cows	26.5	666	21.0	552
	50 to 99	17.4	1,229	15.3	1,077
	100 and over	8.4	1,240	7.7	1,145
	Total	52.3	3,135	44.0	2,774
	Average size of herd	..	60	..	63
	% of total dairy cows in herds of 100 and over	..	39.6	..	41.3
Beef cows	1 to 19 beef cows	50.5	319	48.8	345
	20 to 49	13.6	417	17.3	529
	50 and over	6.4	559	9.0	778
	Total	70.6	1,295	75.2	1,653
	Average size of herd	..	18	..	22
	% of total dairy cows in herds of 50 and over	..	43.2	..	47.1
Breeding sheep	1 to 99 breeding sheep	43.2	1,738	42.1	1,731
	100 to 499	33.4	7,535	37.7	8,587
	500 and over	8.7	8,026	10.9	9,886
	Total	85.3	17,299	90.7	20,205
	Average size of flock	..	203	..	223
	% of total breeding sheep in flocks of 500 and over	..	46.4	..	48.9
Breeding pigs	1 to 49 breeding pigs	12.5	133	8.8	91
	50 to 99	1.8	126	1.3	94
	100 and over	2.4	561	2.3	591
	Total	16.6	819	12.4	776
	Average size of herd	..	49	..	62
	% of total breeding pigs in herds of 100 and over	..	68.5	..	76.2
Fattening pigs	1 to 199 fattening pigs	11.4	505	8.0	347
	200 to 999	3.7	1,722	3.2	1,540
	1,000 and over	1.2	2,567	1.2	2,717
	Total	16.3	4,794	12.4	4,604
	Average size of herd	..	295	..	370
	% of total fattening pigs in herds of 1,000 and over	..	53.6	..	59.0

TABLE 2.4 Numbers and sizes of enterprises (continued)

The data in this table exclude minor holdings

At June of each year

		1986		1991 (provisional) (a)	
		Number of holdings ('000)	Number of livestock ('000)	Number of holdings ('000)	Number of livestock ('000)
Broilers (Includes small numbers of other table fowl in Scotland and Northern Ireland)	1 to 9,999 broilers	1.1	1,409	0.8	1,342
	10,000 to 99,999	0.8	27,850	0.8	30,916
	100,000 and over	0.2	34,038	0.2	43,042
	Total	2.0	63,296	1.8	75,304
	Average size of flock	..	31,444	..	41,127
	% of total broilers in flocks of 100,000 and over	..	53.8	..	57.2
Laying fowls	1 to 4,999 laying fowls	43.4	4,433	32.9	3,077
	5,000 to 19,999	0.8	8,275	0.6	6,083
	20,000 and over	0.4	25,119	0.3	24,189
	Total	44.6	37,827	33.8	33,349
	Average size of flock	..	848	..	986
	% of total laying fowls in flocks of 20,000 and over	..	66.4	..	72.5

(a) The provisional 1991 figures include returns on about 3,500 holdings (net) in Scotland which have been reclassified as main holdings.

TABLE 2.5 Labour force in agriculture

The data cover all holdings (including minor holdings) in England and Wales but exclude minor holdings in Scotland and Northern Ireland (a)

'000 persons   At June of each year

	Average of 1980-82	1987	1988	1989	1990	1991 (provisional)
Workers						
Regular whole-time:						
hired: male	128	98	93	88	85	80
female	11	10	10	11	12	11
family: male	30	30	28	26	25	25
female	5	4	4	4	4	4
Regular part-time:						
hired: male	19	18	19	18	19	18
female	24	22	22	21	21	21
family: male	13	13	13	12	13	13
female	7	7	7	7	7	7
Seasonal or casual:						
male	57	56	56	54	56	55
female	42	38	37	34	35	33
Salaried managers (b)	8	8	8	8	8	8
Total workers	344	304	296	284	283	274
Farmers, partners and directors						
whole-time	206	194	193	189	183	178
part-time	89	90	92	94	98	99
Total farmers, partners and directors	295	284	284	283	282	277
Total farmers, partners, directors and workers (c)	639	588	580	567	565	552
Spouses of farmers, partners and directors (engaged in farm work)	74	77	77	76	77	76
Total labour force (including farmers and their spouses)	714	665	657	644	642	628

(a) Some of the figures shown in this table differ slightly from those in earlier editions of this publication since, for years from 1987 onwards, they now exclude returns on about 2,500 holdings (net) in Scotland which have been reclassified as minor holdings.
(b) The figures for salaried managers relate to Great Britain only.
(c) This is the series referred to as 'Workforce in agriculture' in Table 1.1

TABLE 2.6 Fixed capital stock of agriculture

At end year

	Average of 1980-82	1986	1987	1988	1989	1990
Gross capital stock (£ million, 1985 prices)						
Buildings and works	13,670	15,430	15,480	15,580	15,630	15,650
Plant and machinery	8,410	7,940	7,670	7,450	7,140	6,830
Vehicles	1,590	1,350	1,290	1,260	1,230	1,230
Total	23,670	24,720	24,440	24,290	24,000	23,710
Main types of agricultural machinery ('000 at December of each year, England and Wales only) (a)						
Tractors: under 40 kw	189	152	149	141	137	..
40 kw and over	224	261	266	265	268	..
Tracklaying tractors	11	9	9	9	8	..
Combine harvesters	47	45	...	42	..	..

(a) Data for 1980-82 include machinery owned by agricultural contractors.

3 Policy developments in 1991

European Community developments

1. 1991 was dominated by discussions on CAP reform. In February the Commission produced a paper 'The development and future of the CAP - Reflections Document of the Commission (COM(91)100)' setting out the principles on which, in its view, the Community's future agricultural policy should be based. In July a further paper was published 'The development and future of the Common Agricultural Policy (COM(91)258)' setting out the Commission's ideas for particular commodities in more detail. This was followed by formal proposals for major changes to the regimes for arable crops (cereals, oilseeds, protein crops), tobacco, beef, sheepmeat and milk, together with accompanying proposals for an agri-environmental scheme (replacing the proposals on agriculture in the environment made in October 1990) and for measures on forestry and to encourage early retirement. Negotiations on these proposals are continuing.

2. Against this background, the Commission's price proposals for 1991 did not propose any major changes to existing regimes. However, higher than forecast expenditure in 1991 made it necessary for the Commission to propose a number of economies to ensure that the 1988 Budget Discipline Arrangements and, in particular, the Agricultural Guideline, were respected. Despite pressure from the majority of Member States, who wished to see the Agricultural Guideline increased to take account of the costs of German unification, the package finally agreed achieved this objective. Institutional prices were frozen for most commodities, and reduced for sheepmeat, durum wheat, oilseeds and tobacco; milk quotas were reduced. In addition the cereals co-responsibility levy was increased, although those participating in set-aside schemes will have their levy wholly or partially reimbursed. A further supplement to the annual ewe premium in Less Favoured Areas was agreed.

3. The price fixing also involved a further devaluation of the green pound, to bring it fully into line with the market exchange rate at the time of the settlement. As a result prices in the UK for most commodities were increased by 2 per cent or more in relation to common prices.

4. In January the Agriculture Council agreed to extend the sugar regime unchanged for a further two years. In October the Council reached agreement on a new regime for oilseeds. The new scheme will remain in place pending decisions on the Commission's proposals, in its CAP reform package, for a scheme covering cereals, oilseeds and protein crops.

5. As a result of 1991 production levels, stabiliser mechanisms led to abatements to the common prices set by the Council for: oilseed rape (20.5 per cent), sunflower seed (32 per cent), peas and beans (16.5 per cent), sheepmeat (provisionally 8.1 per cent in Great Britain, 6.45 per cent in Northern Ireland and elsewhere in the Community), cotton (7 per cent), peaches (6 per cent), nectarines (12 per cent), lemons (2 per cent), cauliflowers (3 per cent), satsumas (20 per cent), and oranges (10 per cent). Under the stabiliser arrangements, the additional co-responsibility levy for cereals was set at zero.

6. Following rulings by the European Court of Justice, certain producers who had previously given up milk production for a period (known as SLOM producers) are having quota allocated to them. There was also a very small quota cut for UK milk producers for redistribution to UK producers of additional milk products, such as yogurt, ice-cream and flavoured milk.

GATT Uruguay Round

7. Negotiations on agricultural reform in the Uruguay Round resumed in 1991. Participants agreed to reach "specific, binding commitments" in each of the three main areas under discussion, namely internal support, border measures and export competition. In July, the London Economic Summit confirmed the wish of all those present to complete the negotiations by the end of the year. Discussions continued during the autumn. The United Kingdom continued to play an active role in helping the Commission to formulate a Community position, and in striving to achieve a successful outcome to the Round.

United Kingdom developments

8. A pilot quality wine scheme was launched in June for the 1991 vintage, in line with the Community's existing rules. The UK has successfully pressed the Commission to undertake a study of the possibility of including hybrids deemed suitable for quality wine production for certain North European states, including the UK.

9. A new Code of Good Agricultural Practice for the Protection of Water was launched in July for England and Wales. The Code provides practical guidance to help farmers to avoid causing water pollution, and is one of a number of complementary measures in the Government's action programme to reduce the unacceptably high level of pollution incidents from agricultural practices. The Water (Prevention and Pollution) (Code of Practice) Order came into force in October, giving the Code a statutory basis. Failure to comply with the Code could be taken into account in legal proceedings against a farmer, but would not of itself constitute an offence. A corresponding Code of Good Agricultural Practice for Scotland is at an advanced stage of preparation and in Northern Ireland the Countryside Management Code introduces similar guidance.

10. New regulations controlling the burning of crop residues came into force in July in England and Wales, replacing and extending existing byelaws and the NFU's straw burning code. These measures do not apply in Scotland or Northern Ireland. The new regulations are designed to ensure that

burning is carried out safely and with the minimum degree of nuisance to the general public. A number of exemptions to the total ban on burning, which will come into force in 1993, were announced in September, and include potato haulms, reeds, hop bines, herbage seeds and lavender.

11. A review of the Environmentally Sensitive Areas Scheme in the first areas to be designated concluded that it had successfully achieved its objectives of protecting special landscapes and habitats. In June it was announced that the scheme would be enhanced to secure greater environmental benefits in these areas and in November substantially increased funding was announced for the development of the Scheme throughout the UK. This will include proposals for the designation of six new ESAs in England in 1992 (Lake District, Exmoor, SW Peak, NW Kent Coast, N Dorset/S Wilts Downs, and the Hampshire Avon) and a further six in 1993 (Essex Coast, Upper Thames Tributaries, Cotswold Hills, Blackdown Hills, Shropshire Hills and Dartmoor) and further ESAs in Scotland, Wales and Northern Ireland.

4 Output prices and input costs

Price indices
(Table 4.1)

1. Table 4.1 shows price indices for agricultural products and inputs and Chart 4.1 portrays the main changes over recent years. Between 1990 and 1991 the index of product prices is forecast to fall by just under 2 per cent and that for input prices to rise by 4 per cent. On the product side the index for crop products has risen for the third year running while the index for animals and animal products has fallen for the second year in succession.

TABLE 4.1 Price indices for products and inputs

Indices: 1985=100   Annual average figures for calendar years

	Average of 1980-82	1987	1988	1989	1990	1991 (forecast)
Producer prices for agricultural products	88.8	104.2	103.8	111.8	113.3	111.4
of which:						
Crop products:	91.6	107.7	101.0	107.8	113.8	115.3
Cereals	94.7	98.9	96.2	97.2	98.8	103.7
Root crops	114.3	152.9	121.0	155.9	171.6	169.4
Fresh vegetables	76.5	104.6	100.3	101.9	112.5	110.9
Fresh fruit	78.5	108.3	114.2	120.1	137.0	137.5
Seeds	112.1	124.3	115.2	125.4	131.9	125.9
Flowers and plants	73.3	105.5	110.0	108.3	105.8	113.2
Other crop products	90.7	88.4	78.8	95.7	102.2	89.9
Animals and animal products:	87.5	102.2	105.5	114.2	113.0	109.1
Animals for slaughter	87.9	100.4	103.3	112.0	108.2	101.9
Milk	84.2	107.5	115.2	122.7	123.3	125.8
Eggs	100.4	91.4	77.5	93.2	106.3	91.7
Other animal products	88.3	99.3	98.9	98.0	93.7	90.5
Prices of agricultural inputs	81.6	100.6	105.3	111.7	116.3	120.9
of which:						
Currently consumed in agriculture:	82.5	98.8	103.3	109.3	113.4	117.5
Animal feedingstuffs	87.5	101.3	106.9	112.7	113.6	115.5
Seeds	103.3	100.6	100.7	96.7	102.8	112.0
Animals for rearing and production	67.7	102.6	111.9	116.1	116.5	109.5
Fertilisers and soil improvers	85.0	82.9	85.4	92.4	94.1	91.4
Plant protection products	88.0	106.4	110.4	115.2	126.8	141.1
Maintenance and repair of plant and machinery	71.9	111.0	117.7	124.5	136.2	150.2
Energy, lubricants	70.5	77.7	73.7	81.8	90.7	93.6
Maintenance and repair of buildings	75.6	108.2	113.2	123.4	129.6	133.7
Veterinary services	78.7	106.8	112.3	116.1	122.2	128.5
Materials and small tools	80.0	108.6	114.9	121.6	130.4	140.1
General expenses	75.6	109.4	119.6	126.1	129.2	138.5
Contributing to agricultural investment	77.9	110.0	116.1	124.3	131.4	139.0
Labour costs	74.8	110.2	115.3	124.4	139.1	150.5

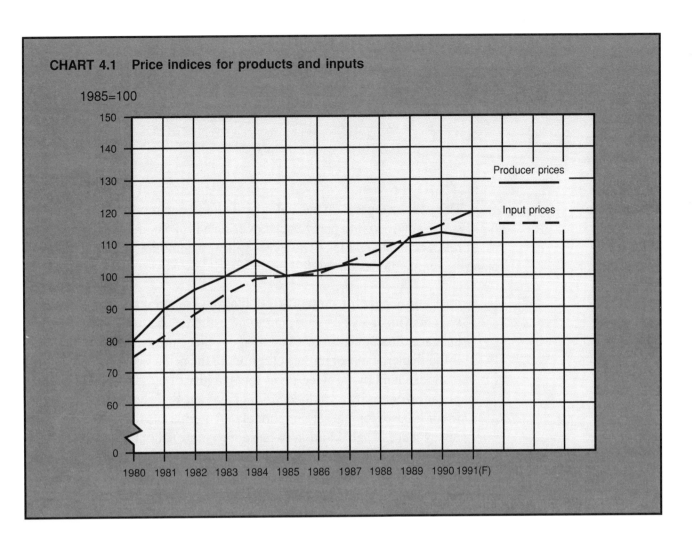

5 Commodities

Introduction

1. This section reports on the major agricultural commodities produced in the UK. It is divided into parts, each covering a broad commodity group, to assist reference to the tables. The tables follow a common format, generally having separate blocks on the following topics:

(i) area and yield (for crops) or populations, marketings and slaughter weights (for livestock), leading to an estimate of production. Allowance for direct on-farm use (on own or other farm but without passing through merchants or millers etc), and for the net increase in the volume of stocks held by farmers, gives the quantity of output (sales). Multiplication of this by a producer price or average realised return (allowing in some cases for market-support related premia and levies and marketing expenses), and addition (when appropriate) of other receipts, gives the value of output of each commodity. These valuations are combined in Table 6.1 in the calculation of the industry's total output and gross and net product and hence in the estimation of incomes from farming. It should however be noted that the valuations of output of each commodity are based on sales within the calendar year and not the quantities produced within the year for sale in that or subsequent years. To assist readers to prepare estimates of the magnitude of each commodity sector on the alternative 'production for sale' basis, Table 6.6 provides details of the value of the changes in on-farm stocks and work-in-progress (and Table 6.5 does the same for breeding livestock capital formation). In the case of input stocks this allows estimation of the value of the usage of (purchased) inputs within the year as well as expenditure on them;

(ii) the sources of new supplies and, in some cases, their various uses. Total new supply is defined as production plus imports less exports. This total new supply may be augmented (or reduced) by a decrease (or increase) in stocks. The result after allowing for these changes is 'total domestic uses'; for the major cereals and milk the most important uses are identified separately;

(iii) home production as a percentage of total new supply and in some cases the level of closing stocks.

Cereals

Cereals *(Tables 5.1-5.6)*

2. In 1991 cereal production was estimated at 22.7 million tonnes, very similar to total production in 1990 and 1989. Higher yields offset a reduction in the total area sown to cereals. The wheat area was slightly down on 1990, with an increase in plantings of higher yielding feed wheats at the expense of

breadmaking wheat varieties. The barley area continued to decline, being about 8 per cent down on 1990. The overall quality of the harvest was good though quality was more variable than in recent years.

3. The EC 1990 harvest was estimated at 159.7 million tonnes, excluding production from the former German Democratic Republic. As this was below the Maximum Guaranteed Quantity (MGQ), no price cuts or additional co-responsibility levy were triggered under the stabiliser rules. Preliminary indications of the 1991 EC harvest are that it will exceed the MGQ.

4. In the price fixing, the basic cereals co-responsibility levy was increased to 5 per cent, but it was also agreed that those participating in a new one-year set-aside scheme should be reimbursed levy paid on cereals during the year to 30 June 1992. Those participating in the five-year scheme will also be entitled to refund of a 2 per cent levy on grain marketed in the same period.

5. In the first half of 1991 ex-farm prices of feed wheat, feed barley and breadmaking wheat were considerably higher than in the corresponding period of 1990. In the second half of the year ex-farm prices of new crop feed wheat and barley were closer to 1990 levels. Premium malting barley prices were generally slightly down on the previous year but breadwheat prices were some £10-£15 per tonne higher, reflecting the reduced availability of breadmaking wheat supplies.

TABLE 5.1 Wheat

'000 tonnes (unless otherwise specified)   Calendar years

	Average of 1980-82	1987	1988	1989	1990	1991 (forecast)
Area, yield, production and output						
Area ('000 hectares)	1,532	1,994	1,886	2,083	2,013	1,990
Yield (tonnes/hectare)	5.98	5.99	6.23	6.74	6.97	7.20
Production	9,163	11,940	11,751	14,033	14,033	14,333
Direct use on farms	..	1097	954	964	999	1,014
Increase in farmers' stocks	..	-396	-185	20	143	66
Output	8583	11,238	10,981	13,048	12,893	13,253
Average realised price (£/tonne)	107.9	112.5	105.2	104.9	109.9	116.6
Value of output (£ million)	926	1,264	1,155	1,368	1,417	1,546
Supply and utilisation						
Production	9,163	11,940	11,751	14,033	14,033	14,333
Imports from: the Eleven	284	1,233	1,137	520	523	434
third countries	1,585	479	504	339	350	400
Exports to: the Eleven	974	1,593	1,085	1,301	1,859	2,357
third countries	711	2,525	843	1,978	2,777	1362
Total new supply	9,347	9,534	11,464	11,611	10,270	11,448
Increase in farm and other stocks	170	-2,258	-606	121	-840	768
Total domestic uses	9,177	11,792	12,070	11,490	11,110	10,680
of which: flour milling	4,726	4,919	5,111	5,032	4,878	4,819
animal feed	3,868	5,817	5,796	5,258	5,017	4,591
seed	287	337	357	365	353	347
other uses and waste	296	719	806	835	862	923
Production as % of total new supply for use in UK	98	125	103	121	137	125
% of home grown wheat in milling grist	64.2	77.8	65.6	82.7	87.6	86.6

TABLE 5.2 Barley

'000 tonnes (unless otherwise specified) Calendar years

	Average of 1980-82	1987	1988	1989	1990	1991 (forecast)
Area, yield, production and output						
Area ('000 hectares)	2,294	1,832	1,880	1,653	1,517	1,401
Yield (tonnes/hectare)	4.58	5.04	4.67	4.88	5.21	5.50
Production	10,502	9,229	8,778	8,073	7,897	7,707
Direct use on farms	..	2,785	2,770	2,604	2,521	2,281
Increase in farmers' stocks	..	-114	-33	-594	-133	14
Output	7,774	6,560	6,041	6,063	5,509	5,413
Average realised price (£/tonne)	101.1	105.5	103.7	106.7	108.8	108.9
Value of output (£ million)	786	692	627	647	599	589
Supply and utilisation						
Production	10,502	9,229	8,778	8,073	7,897	7,707
Imports from: the Eleven	121	236	312	277	255	103
third countries	5	...	6	...	...	120
Exports to: the Eleven	1,191	1,048	731	427	916	950
third countries	1,188	2,047	2,079	2,816	1,106	1,043
Total new supply	8,250	6,369	6,285	5,107	6,130	5,937
Increase in farm and other stocks	35	-233	-141	-1,177	-13	-249
Total domestic uses	8,215	6,602	6,426	6,285	6,143	6,186
of which: brewing/distilling	1,931	1,740	1,745	1,851	1,835	1,956
animal feed	5,722	4,428	4,241	4,017	3,939	3,876
seed	361	256	273	243	220	206
other uses and waste	201	178	167	174	149	148
Production as % of total new supply for use in UK	127	145	140	158	129	130

TABLE 5.3 Oats

'000 tonnes (unless otherwise specified)   Calendar years

	Average of 1980-82	1987	1988	1989	1990	1991 (forecast)
Area, yield, production and output						
Area ('000 hectares)	140	99	120	119	107	105
Yield (tonnes/hectare)	4.27	4.57	4.55	4.46	4.96	5.18
Production	598	454	548	529	530	545
Direct use on farms	..	198	232	178	216	216
Increase in farmers' stocks	..	2	22	33	-4	0
Output	260	254	295	318	318	329
Average realised price (£/tonne)	98.7	115.4	106.5	98.7	106.1	107.4
Value of output (£ million)	26	29	31	31	34	35
Supply and utilisation						
Production	598	454	548	529	530	545
Imports from: the Eleven	8	7	8	8	10	7
third countries	3	...	1	...	1	1
Exports to: the Eleven	2	9	5	14	8	12
third countries	...	...	...	...	...	...
Total new supply	607	453	551	523	532	541
Increase in farm and other stocks	-13	-3	36	34	-16	-4
Total domestic uses	620	456	515	489	548	545
of which: milling	143	162	185	225	235	227
animal feed	433	261	288	227	275	281
seed	29	20	27	21	22	20
other uses and waste	15	13	15	16	16	17
Production as % of total new supply for use in UK	98	100	99	101	100	101

Table 5.4 Rye, Mixed Corn and Triticale

'000 tonnes (unless otherwise specified)   Calendar years

	Average of 1980-82	1987	1988	1989	1990	1991 (forecast)
Area, production and output (a)						
Area ('000 hectares)	18	13	19	20	22	24
Production	72	58	85	94	109	129
Value of output (£ million)	2	3	3	3	4	4
Supply and utilisation						
Imports from (b): the Eleven	17	19	26	13	12	12
third countries	...	1	...	...	1	2
Exports to (b): the Eleven	...	...	...	...	5	...
third countries	...	...	...	...	...	...
Total new supply	89	78	111	107	117	143
Production as % of total new supply for use in UK	81	74	77	88	93	90

(a) Triticale has been included from 1988 onwards, with the area figure for 1988 and the production figures for 1988 and 1989 being estimates.
(b) Relates only to rye and triticale.

Table 5.5 Maize

'000 tonnes Calendar years

	Average of 1980-82	1987	1988	1989	1990	1991 (forecast)
Supply and utilisation						
Imports from: the Eleven	774	1,207	1,156	1,273	1,407	618
third countries	1,623	303	191	164	239	1,113
Exports to: the Eleven	10	20	7	26	15	14
third countries	1	...	...	...	...	...
Total new supply	2,386	1,490	1,340	1,411	1,631	1,717

Table 5.6 Total cereals

'000 tonnes (unless otherwise specified) Calendar years

	Average of 1980-82	1987	1988	1989	1990	1991 (forecast)
Area, production and output						
Area ('000 hectares)	3,983	3,937	3,905	3,874	3,658	3,520
Production	20,335	21,681	21,162	22,729	22,569	22,713
Output	16,641	18,079	17,343	19,458	18,757	19,029
Supply and utilisation						
Imports from: the Eleven	1,204	2,702	2,639	2,091	2,207	1,173
third countries	3,216	783	702	503	591	1,636
Exports to: the Eleven	2,177	2,670	1,828	1,768	2,803	3,333
third countries	1,900	4,572	2,922	4,794	3,883	2,405
Total new supply	20,678	17,924	19,753	18,761	18,681	19,784
Increase in stocks	185	-2,479	-711	-1,024	-869	515
Total domestic uses	20,493	20,403	20,462	19,783	19,550	19,273
Production as % of total new supply for use in UK	99	121	107	121	121	115
Stocks (of wheat, barley and oats) held at year end: by farmers (a)	..	9,268	9,070	8,530	8,535	8,959
in intervention (b)	..	1,729	1,544	705	633	766
by processors and traders (a)	..	5,268	4,938	5,293	4,491	4,658

(a) Stocks held by agricultural co-operatives have been included in processors' and traders' stocks prior to 1987, and in farmers' stocks from 1987.
(b) Recorded as all physical stocks held at the year end.

Other crops

Oilseed rape
(Table 5.7)

6. There was a significant increase in the area planted to oilseed rape, from 390,000 hectares in 1990 to an estimated 445,000 hectares. Production is forecast at 1.31 million tonnes, a small increase over 1990 production (1.26 million tonnes) but below the 1987 peak of 1.35 million tonnes.

7. Community production (EC10) has risen from 5.82 million tonnes in 1990 to an estimated 6.38 million tonnes, an increase of over 9 per cent. Under the stabiliser mechanism, the baseline target price has been cut by 20.5 per cent for the 1991/92 marketing year compared with a 15.5 per cent cut applied in 1990/91. Taking into account the 1.5 per cent reduction in the target price agreed at the price-fixing and the green pound devaluation of 1 July 1991, the support price in the UK for 1991/92 is reduced by 5.4 per cent (£16.10/tonne) compared with the 1990 level.

8. A new Community support scheme for the major oilseeds (including rapeseed) has been agreed, to take effect from 1 July 1992. Support will take the form of direct per hectare payments to producers, rather than tonnage payments to processors as under the present system. The new scheme will remain in place pending decisions on Commission proposals for a scheme covering cereals, oilseeds and protein crops.

Sugar beet and sugar
(Table 5.8)

9. The area of sugar beet harvested in 1991 is expected to be around 170,000 hectares. Most of the crop was sown in good conditions by mid-April. Subsequent dry weather has constrained yields but the crop withstood the drought well. No incidence of rhizomania was detected in sugar beet crops. White sugar production is forecast to be about 1.27 million tonnes, compared with 1.24 million tonnes in 1990. 72,000 tonnes of white sugar which were produced in excess of quota in 1990 were carried forward into 1991. Total supply is expected to be significantly above the maximum quota of 1.144 million tonnes set for the UK under the European Community sugar regime. The excess has to be exported without export refunds or carried forward to count against next year's quota.

Hops *(Table 5.9)*

10. The indications are that yields and alpha acid contents are higher than in the 1990 crop, and that there is a slight increase in the prices for hops sold under contract.

Peas and beans
(Table 5.10)

11. Production of peas for harvesting dry, including those for human consumption, has decreased from 320,000 tonnes in 1990 to an estimated 262,000 tonnes. The decrease results from lower forecast yields and a fall in area sown from 77,000 to an estimated 74,000 hectares. Production of beans fell from 474,000 tonnes in 1990 to an estimated 437,000 tonnes. The area sown decreased from 139,000 hectares in 1990 to an estimated 129,000 hectares.

12. Community peas and beans production is estimated to have fallen to 4.465 million tonnes in 1991 from 4.83 million tonnes in 1990. Under the

stabiliser mechanism, the baseline guide price has been reduced by 16.5 per cent for the 1991/92 marketing year compared with a 20 per cent cut applied in 1990/91. Taking into account the green pound devaluation of 1 July 1991, the support price in the UK for 1991/92 is 4.9 per cent (£8.99/tonne) higher than the 1990 level.

Seeds
(Tables 5.12 and 5.13)

13. Total production of herbage and legume seeds in the year ended May 1991 fell by 9 per cent on the 18,600 tonnes produced in the previous year. At 7,100 tonnes, imports of seeds from other Member States of the European Community during 1990/91 showed a decrease of 23 per cent. The area approved for the production of herbage and legume seed in 1991/92 is expected to rise by 8.6 per cent from the 1990/91 level.

Potatoes *(Table 5.14)*

14. UK imports of ware potatoes in the first half of 1991 were approximately half those for the comparable period of 1990. Frost in April followed by prolonged periods of unseasonably cool and then dry weather delayed the start of the harvesting of the 1991 crop by about a month. Harvesting began slowly and there was no return to the very low prices which had led to the Potato Marketing Board intervening in the market in the previous year. Although stored potatoes were available they were of variable quality and this, together with a late harvest, meant that there was a substantial increase in the import of new potatoes during July 1991.

15. Plantings in the UK in 1991 were estimated at 177,400 hectares, about 500 hectares less than the previous year. Yields in both 1990 and 1991, which were both dry years, were down on previous years. Yields of early potatoes were significantly lower in 1991 than in 1990, and those for the maincrop fell slightly.

16. The continuation of the Potato Marketing Scheme will be kept under review. It remains the Government's intention to abolish the guarantee arrangements as soon as Parliamentary time permits but, in the meantime, the industry is now bearing a greater share of the cost of its price support arrangements. In Northern Ireland, market support arrangements are provided for in an agreement between the Department of Agriculture and the Ulster Farmers' Union.

TABLE 5.7 Oilseed rape

'000 tonnes (unless otherwise specified)   Calendar years

	Average of 1980-82	1987	1988	1989	1990	1991 (forecast)
Area ('000 hectares)	130	388	347	321	390	445
Yield (tonnes/hectare) (a)	3.12	3.49	3.00	3.04	3.23	2.94
Production (a)	407	1,353	1,040	976	1,258	1,308
Average producer price (b) (£/tonne)	256	221	232	286	273	248
Value of output (£ million)	104	298	241	279	343	324
Imports from: the Eleven	64	151	67	71	117	126
third countries	41	40	14	31	94	...
Exports to: the Eleven	5	295	150	103	193	125
third countries	...	1	...	...	...	...
Total new supply	506	1,248	971	975	1,276	1,309
Production as % of total new supply for use in UK	80	108	107	100	99	100

(a) These figures are on the basis of a standardised (9%) moisture content.
(b) Received by farmers for the year's crop, including 'Double-low' varieties. An adjustment is made for drying costs.

TABLE 5.8 Sugar beet and sugar

'000 tonnes (unless otherwise specified)   Calendar years

	Average of 1980-82	1987	1988	1989	1990	1991 (forecast)
Sugar beet						
Area ('000 ha) (a)	209	202	201	197	194	196
Yield (tonnes/ha) (a)	39.57	39.47	40.65	41.28	40.66	40.32
Production of beet	8,261	7,992	8,152	8,113	7,902	7,900
Average market price (b) (£/tonne)	25.79	27.85	29.27	30.83	34.43	34.00
Value of output (£ million)	213	223	239	250	272	269
Sugar content %	16.58	16.78	17.65	17.22	17.50	17.60
Sugar extraction rate %	88.31	91.0	91.0	91.0	91.0	91.0
Sugar ('000 tonnes refined basis)						
Production (c)	1206	1,226	1,304	1,267	1,241	1,270
Imports from (d): the Eleven (e)	163	93	121	128	154	137
third countries	1093	1,084	1,178	1,195	1,138	897
Exports to (d): the Eleven	10	40	64	78	37	41
third countries	111	288	227	277	292	177
Total new supply	2341	2,075	2,312	2,235	2,204	2,086
Production as % of total new supply for use in UK	51	59	56	57	56	61

(a) The area, and related yield figures, are based on June census definitions. They differ from those for the harvested area (and the related yields), which figures (following a recent survey) are estimated to have been 170,000 ha and 46.47 tonnes/ha in 1991.
(b) Estimated as the return to grower price less transport costs.
(c) Sugar coming out of the factory in the early part of the new year is regarded as being part of the production of the previous calendar year.
(d) Includes only sugar as such and takes no account of the sugar content of processed products.
(e) Includes imports from French overseas departments.

TABLE 5.9 Hops

'000 tonnes (unless otherwise specified)   Calendar years

	Average of 1980-82	1987	1988	1989	1990	1991 (forecast)
Area ('000 hectares) (a)	5.8	4.2	4.0	3.9	3.9	3.8
Yield (tonnes/hectare)	1.68	1.23	1.22	1.20	1.18	1.41
Production	9.8	5.2	4.9	4.7	4.6	5.3
Farm gate price (£/tonne)	2,582	2,358	2,468	2,662	2,922	3,303
Total realised return (£ million)	25	12	12	13	13	18
Other receipts (£ million) (b)	1	1	1	2	1	1
Value of production (£ million)	26	13	13	14	14	18
Imports from: the Eleven	1.3	1.2	1.3	1.4	0.8	0.7
third countries	0.6	1.0	0.9	0.9	0.8	0.6
Exports to: the Eleven	2.9	0.8	0.8	2.1	1.4	1.0
third countries	0.7	0.1	0.1	0.1	...	0.1
Total new supply	8.2	6.5	6.3	4.8	4.6	5.5
Production as % of total new supply for use in UK	120	80	79	98	99	96

(a) The area is that recorded in the June Census (and shown in Table 2.2), not all of which may actually be in production within the year. The yield relates to the Census area.
(b) Production aid.

TABLE 5.10 Peas and beans for stockfeed

Calendar years

	Average of 1980-82	1987	1988	1989	1990	1991 (forecast)
Peas for harvesting dry (a)						
Area ('000 hectares)	..	99.1	90.6	72.3	65.2	62.6
Yield (tonnes/hectare)	..	2.65	3.38	3.25	4.17	3.55
Production ('000 tonnes)	..	262.6	306.3	235.5	272.2	222.2
Price (£/tonne)	..	198.3	160.4	172.7	160.1	169.9
Value of production (£ million)	..	52	49	41	44	38
Field beans (mainly for stockfeed)						
Area ('000 hectares)	..	91.0	153.7	129.0	139.2	129.1
Yield (tonnes/hectare)	..	3.23	3.88	3.49	3.41	3.38
Production ('000 tonnes)	..	293.9	596.4	450.2	474.0	436.7
Price (£/tonne)	..	193.0	156.2	168.4	164.0	165.8
Value of production (£ million)	..	57	93	76	78	72

(a) The figures presented here cover only that part of the crop which is assumed to be used for stockfeed; the remainder is included in Horticulture, Table 5.15.

TABLE 5.11 Purchased feedingstuffs (excluding direct inter-farm sales)

Million tonnes (unless otherwise specified)   Calendar years

	Average of 1980-82	1987	1988	1989	1990	1991 (forecast)
Concentrates						
Compounds for:						
cattle	4.7	3.8	3.7	3.8	3.8	3.7
calves	0.4	0.4	0.3	0.3	0.3	0.3
pigs	2.2	2.2	2.2	2.1	2.3	2.4
poultry	3.4	3.5	3.6	3.4	3.7	3.8
other	0.4	0.6	0.7	0.7	0.8	0.8
Total	11.2	10.5	10.5	10.4	10.9	11.0
Straight concentrates (ie cereals, cereal offals, proteins and other high energy feeds)	3.5	4.8	4.7	4.2	4.0	4.3
Total	14.7	15.3	15.2	14.7	14.9	15.3
Non - concentrates (low-energy bulk feeds expressed as concentrate equivalent) (a)	0.7	0.7	0.6	0.7	0.7	0.7
Total all purchased feedingstuffs	15.3	15.9	15.9	15.3	15.6	16.0
Cost of purchased feedingstuffs (£ million)	2,248	2,614	2,731	2,747	2,848	2,957
of which this part represents the value of sales off the national farm (and included in output) but subsequently repurchased as an input (£ million)	626	566	575	553	478	586

(a) Brewers and distillers grains, hay, straw, milk by-products and other low-energy bulk feeds expressed in terms of equivalent tonnage of high energy feeds.

TABLE 5.12 Herbage and legume seeds (excluding field bean and field pea seeds)

'000 tonnes (unless otherwise specified) June/May years

	Average of 1980/81-1982/83	1987/88	1988/89	1989/90	1990/91	1991/92 (forecast)
Certified seed:						
area ('000 hectares)	16.7	17.8	17.7	17.4	14.6	15.9
production	16.6	15.9	16.6	18.6	17.0	15.5
Imports from: the Eleven	8.1	6.9	5.3	9.2	7.1	..
third countries	5.6	3.9	4.5	3.9	4.8	..
Exports to: the Eleven	4.1	3.1	2.3	2.3	1.3	..
third countries	0.6	0.7	0.3	0.3	0.6	..
Total new supply	25.6	22.9	23.8	29.1	29.6	..

TABLE 5.13 Purchased seeds (excluding direct inter-farm sales)

'000 tonnes Calendar years

	Average of 1980-82	1987	1988	1989	1990	1991 (forecast)
Cereals	603.7	550.4	582.8	555.0	518.6	497.0
Grass and clover	20.4	14.7	14.7	17.2	14.2	14.1
Root and fodder crops	18.9	51.4	52.8	43.8	45.3	42.6
Potatoes	304.0	251.0	265.0	269.0	278.0	266.0
Vegetable and other horticultural seeds (a)	21.5	21.4	21.0	20.9	20.9	19.4
Total cost of all purchased seeds (£ million)	222	282	280	291	298	293
of which this part represents the value of sales off the national farm (and included in output) but subsequently repurchased as an input (£ million).	104	133	131	137	140	138

(a) Includes mushroom spawn, bulbs and seeds for hardy nursery stock, flowers, sugar beet and oilseed rape.

TABLE 5.14 Potatoes

'000 tonnes (unless otherwise specified)   Calendar years

	Average of 1980-82	1987	1988	1989	1990	1991 (forecast)
Area, yield, production and output						
Area ('000 hectares): early	20.3	16.6	17.5	17.5	15.9	16.4
maincrop	176.7	162.0	163.7	157.6	162.0	160.9
Yield (tonnes/hectare): early	20.8	23.6	24.0	20.9	27.4	21.9
maincrop	35.5	39.0	39.6	37.4	37.3	37.2
Production: early	421	392	420	367	437	360
maincrop	6,272	6,322	6,479	5,895	6,043	5,994
total	6,693	6,713	6,899	6,262	6,480	6,354
Waste	753	794	901	798	680	748
Increase in farmers' stocks	-85	102	69	-434	25	-221
Total output	6,025	5,817	5,929	5,898	5,775	5,827
Average price (£/tonne) (a) paid to registered producers for: early potatoes	87.6	119.4	103.2	132.6	82.7	142.9
maincrop potatoes	61.8	82.4	64.5	80.3	89.3	85.8
realised for all potatoes	63.7	83.5	67.3	82.9	87.9	87.4
Value of output of potatoes (£ million)	383	486	399	489	508	509
Supplies and utilisation						
Total production	6,693	6,713	6,899	6,262	6,480	6,354
Supplies from the Channel Islands	31	43	35	31	41	38
Imports	710	985	947	1,005	897	842
of which: early from: the Eleven	47	59	40	84	55	49
third countries	258	133	148	154	177	166
maincrop from: the Eleven	160	372	230	195	104	85
third countries	9	7	14	...	10	4
processed (raw equivalent) from: the Eleven	175	375	484	527	480	462
third countries	50	19	4	16	41	46
seed from: the Eleven	9	21	26	30	30	30
third countries	1	...	...	...	...	...
Exports	213	266	164	207	224	198
of which: raw to: the Eleven	20	39	32	60	52	33
third countries	34	55	35	40	48	41
processed (raw equivalent) to: the Eleven	33	52	36	43	53	48
third countries	6	7	2	3	7	1
seed to: the Eleven	3	52	35	36	26	33
third countries	118	61	24	26	39	41
Total new supply	7,220	7,475	7,717	7,091	7,193	7,036
Farmers' stocks: opening stocks	3,725	2,640	2,742	2,811	2,376	2,401
closing stocks	3,640	2,742	2,811	2,376	2,401	2,180
net increase	-85	102	69	-434	25	-221
Total domestic uses	7,305	7,373	7,648	7,526	7,169	7,256
of which: used for human consumption	5,772	5,987	6,022	6,033	5,903	5,860
seed for home crops (including seed imports)	594	563	568	563	573	573
support buying	147	...	142	92	12	17
chats, waste and retained stockfeed	792	823	916	837	680	806
Production as % of total new supply for use in UK	93	90	89	88	90	90

(a) Including sacks.

Horticulture

Horticulture
(Tables 5.15 and 5.16)

17. The total area devoted to horticulture as recorded in the June agricultural census was 200,000 hectares in 1991 (provisional) compared to 208,000 hectares in 1990. However, the area figures shown in Table 5.15 differ from these for several reasons, including multi-cropping. This table gives information about the main sectors. More detailed information is given in Table 5.16 for cauliflowers, tomatoes, apples and pears, the four crops for which intervention arrangements apply.

18. The cooler summer resulted in poorer returns for salad crops. The area of protected lettuce declined although that for field lettuce increased, and prices were lower. Competition from the continent for Iceberg was strong. Average yields of tomatoes were obtained and early season prices were again poor, with only a modest improvement in mid season. Returns for cucumbers were lower than in 1990.

19. The area devoted to field vegetables has shown little change on last year. Dry weather later in the season again affected supplies and returns were reduced largely due to lower returns for carrots, onions and lettuce. The changeable weather also caused irregular summer cauliflower production. Dwarf beans were one of the few crops which showed upturns in output, price and value. The overall decline in Brussels sprouts has continued.

20. The total area of top fruit continued to decline in 1991 but production was up slightly compared with 1990. Low temperatures in May and June followed by dry conditions in August and September resulted in a greater proportion of smaller fruit. Apple withdrawals decreased in 1990/91 and are likely to be very low in 1991/92 due to the strong demand on the UK market and a poorer continental crop. Yields of Cox, Bramley's Seedling and Conference pears were slightly higher than last year.

21. In the soft fruit sector, wastage of strawberries was 50 per cent up on last year resulting in lower marketable output, and prices were lower than in 1990. Late season crops of strawberries and raspberries were variable due to disease problems related to climatic conditions. In Scotland, planted areas fell significantly, reflecting the effect of East European imports. The blackcurrant crop was larger than in 1990 but prices were lower.

22. Within the ornamental sector the overall market was less buoyant than in previous years, although the bedding plants sector continued to expand. Good returns were realised in the spring for hardy nursery stock and bedding plants. The 1991 season produced good flower bulb yields but the market continued to be oversupplied. Production of outdoor cut flowers continued to increase.

TABLE 5.15   Horticulture

Calendar years

	Average of 1980-82	1987	1988	1989	1990	1991 (forecast)
Vegetables						
Grown in the open (a):						
area ('000 hectares) (b)	187.0	190.6	195.5	192.1	191.3	192.0
value of output (£ million)	371	565	579	619	636	617
Protected:						
area ('000 hectares) (b) (c)	2.8	2.7	2.8	2.7	2.7	2.6
value of output (£ million)	177	311	321	321	358	325
Total value of output (£ million)	549	876	900	939	994	941
Fruit						
Orchard fruit:						
area ('000 hectares) (b)	41.5	34.9	34.1	33.0	32.2	32.1
value of output (£ million)	100	114	119	149	159	157
Soft fruit:						
area ('000 hectares) (b)	18.4	14.5	14.5	14.5	14.4	14.1
value of output (£ million)	74	108	115	112	114	117
Total value of output (£ million) (d)	174	223	234	261	273	274
Ornamentals						
Area ('000 hectares) (b)	13.0	17.4	18.6	18.3	18.6	18.6
Value of output (£ million)	191	320	389	446	495	503
of which: flower bulbs	23	26	26	30	33	33
flowers in the open	14	18	21	23	25	27
hardy nursery stock	85	143	182	213	237	236
protected crops	70	133	160	180	200	207
Seeds: value of output (£ million)	4	9	6	7	6	4
Total value of commercial horticultural output (£ million) (e)	919	1,428	1,529	1,654	1,768	1723
Value of output of main crops (£ million)						
cabbages	50	80	83	87	88	85
carrots	38	69	83	81	84	79
cauliflowers	56	60	62	60	64	64
lettuces	48	101	90	98	105	93
mushrooms	81	145	162	166	175	169
peas (a)	53	47	56	58	63	59
tomatoes	51	81	74	73	90	78
apples	71	84	86	115	126	120
pears	12	14	15	16	18	19
raspberries	19	29	34	32	24	30
strawberries	42	63	63	62	67	60

(a) Includes peas harvested dry for human consumption.
(b) Areas relate to actual cropped areas which can differ from Census areas (Table 2.2).
(c) Excludes mushrooms area.
(d) Includes glasshouse fruit.
(e) Includes hedgerow fruit and nuts

TABLE 5.16 Selected horticultural crops

'000 tonnes (unless otherwise specified) — Calendar years

	Average of 1980-82	1987	1988	1989	1990	1991 (forecast)
Cauliflowers						
Farm gate price (£/tonne)	181.6	211.4	182.3	193.1	221.9	214.9
Output	310	282	341	313	288	299
Supplies from Channel Islands	12	10	11	7	7	8
Imports from: the Eleven	35	41	60	53	33	38
third countries	...	...	...	...	...	...
Total new supply	357	333	412	373	328	345
Output as % of total new supply for use in UK	87	85	83	84	88	87
Tomatoes						
Farm gate price (£/tonne)	418.7	671.6	573.5	555.9	673.4	595.6
Output	121	121	129	132	134	131
Supplies from Channel Islands	48	18	16	17	13	16
Imports from: the Eleven	76	152	166	175	163	152
third countries	120	104	100	98	90	85
Exports to: the Eleven	5	6	7	6	5	5
third countries	...	...	...	...	...	...
Total new supply	360	389	404	416	394	379
Output as % of total new supply for use in UK	34	31	32	32	34	35
Apples (excluding cider apples)						
Farm gate price (£/tonne):						
dessert	278.6	335.6	412.0	363.8	495.7	538.8
culinary	218.3	244.0	287.1	210.4	280.4	320.4
Output from the crop:						
dessert	167	160	132	202	179	143
culinary	118	124	111	197	134	135
Imports from: the Eleven	264	291	312	274	274	277
third countries	135	152	187	196	193	211
Exports to: the Eleven	14	32	16	31	20	25
third countries	1	...	...	...	...	...
Total new supply	668	695	726	839	760	740
Increase in stocks	-6	-20	7	15	-26	35
Total domestic uses	662	675	733	854	734	775
Output as % of total new supply for use in UK	43	41	33	48	41	38
Closing stocks	89	93	100	115	89	124

TABLE 5.16 Selected horticultural crops (continued)

'000 tonnes (unless otherwise specified)   Calendar years

	Average of 1980-82	1987	1988	1989	1990	1991 (forecast)
Pears (excluding perry pears)						
Farm gate price (£/tonne)	274.2	308.1	331.6	424.0	525.1	548.0
Output from the crop	46	45	44	37	34	35
Imports from: the Eleven	50	55	63	73	65	66
third countries	17	22	30	29	33	33
Exports to: the Eleven	1	1	1	1	2	2
third countries	...	...	...	...	...	...
Total new supply	112	121	136	138	130	132
Increase in stocks	4	-13	...	4	-1	2
Total domestic uses	115	108	136	141	129	133
Output as % of total new supply for use in UK	42	37	32	27	26	27
Closing stocks	18	13	13	16	16	17

Livestock

Cattle and calves; beef and veal
(Table 5.17)

23. The size of the beef breeding herd continued to increase, reaching its highest level for twelve years, with the increase in beef cows partially offsetting the further decrease of the dairy herd. For the first four months of 1991 average market prices were below last year's levels. Prices from May to October, though showing the usual seasonal decline, were above those for the same period last year.

24. During the first quarter of 1991, 38,000 tonnes of beef were accepted into intervention. In the context of the 1991 price fixing agreement, measures were introduced to weaken the intervention system in the hope of making it less attractive to operators. Despite these measures, intervention buying continued. In the UK, purchases from April (the start of the intervention year) to November remained high with some 82,000 tonnes of beef being bought in compared with 85,000 tonnes over the same period in 1990. Although sales from intervention remained limited as a result of the contraction of the Middle East market and continued concern in importing countries over BSE, exports of non-intervention beef recovered well. Furthermore, domestic household consumption of beef in the third quarter of 1991 moved above the equivalent level for 1990. Producers continued to benefit from payments under the beef special premium and suckler cow premium schemes.

Sheep and lambs; mutton and lamb
(Table 5.18)

25. The size of the breeding flock has remained largely constant over the past year, bringing to an end a long period of expansion. Domestic production of sheepmeat in 1991 is estimated to be about 6 per cent higher than in 1990. Exports are expected to be around 2 per cent higher than last year whilst supplies to the home market are expected to be 2 per cent below those of 1990. Average market prices for Great Britain are expected to be around 18 per cent below those of last year. Average market prices for Northern Ireland are expected to increase by around 5 per cent over 1990.

26. The budgetary stabiliser mechanism for sheepmeat this year gave rise to a provisional cut of 8.1 per cent applied to the 1991 guide price, which is used for determining the variable premium in Great Britain. The 1990 definitive stabiliser reduction, applied to the basic price for calculating the 1990 annual ewe premium, was 7 per cent for both Great Britain and Northern Ireland. The Sheep Variable Premium Scheme terminated at the end of the 1991 marketing year.

27. The Agriculture Departments continue to pay compensation to sheep producers whose activities have been affected by restrictions imposed following the Chernobyl accident in 1986. Compensation by the end of 1991 is expected to be in excess of £8.8 million.

Pigs and pigmeat
(Table 5.19)

28. Producer prices were low at the start of the year but recovered following the introduction of a Community wide private storage aid scheme. Prices were again severely depressed during the summer because of poor demand and competition from cheaper imports and from lamb and poultry meat, before a

limited recovery towards the end of the year.

29. The incidence of Blue Ear Disease in the North East of England had a minimal impact on supplies of pigmeat which showed little change over the previous year.

Poultry and poultrymeat
(Table 5.20)

30. Following the fall-back in 1989 and 1990, UK production of chicken returned to 1988 levels. However, with imports continuing at a relatively high level, prices to producers remained low throughout the year while stocks increased. Turkey production has increased and has almost recovered to 1989 levels.

TABLE 5.17 Cattle and calves; beef and veal

Calendar years

	Average of 1980-82	1987(a)	1988	1989	1990	1991 (forecast)
Populations						
Total cattle and calves ('000 head at June)	13,269	12,170	11,884	11,975	12,059	11,871
of which: dairy cows	3,223	3,042	2,912	2,865	2,847	2,765
beef cows	1,429	1,345	1,375	1,495	1,599	1,670
dairy heifers in-calf	688	599	612	566	529	539
beef heifers in-calf	162	176	222	227	227	195
other	7,767	7,008	6,763	6,822	6,857	6,701
Selected market prices						
Store cattle (£ per head) (b):						
1st quality Hereford/cross bull calves (c)	75	134	186	188	128	118
1st quality beef/cross yearling steers (d)	186	360	414	431	397	390
Finished cattle (p per kg liveweight): All clean cattle	80.2	96.1	109.2	113.9	106.2	106.9
Marketings, production and returns						
Total home-fed marketings ('000 head)	4,196	4,463	3,598	3,710	3,835	4,013
of which: steers, heifers and young bulls	2,841	3,075	2,619	2,664	2,774	2,792
calves	373	459	286	350	400	483
cows and adult bulls	982	929	693	697	661	738
Average dressed carcase weights (dcw) (kgs) (e):						
steers, heifers and young bulls	269.8	276.0	282.9	288.0	288.8	286.6
calves	43.3	51.3	55.2	49.6	40.8	39.0
cows and adult bulls	264.0	272.9	276.7	277.6	278.9	279.0
Production ('000 tonnes, dcw):						
Home-fed production	1,039	1,119	944	972	998	1,021
Gross indigenous production (f)	..	1,103	923	959	986	1,010
Average realised return (p per kg dcw) (g)	153	177	200	202	181	182
Total realised return (£ million)	1,594	1,981	1,889	1,962	1,806	1,854
Other receipts (£ million) (h)	54	88	87	139	167	181
Value of home-fed production (£ million)	1,648	2,069	1,976	2,100	1,973	2,035
Supplies ('000 tonnes, dcw) (i)						
Home-fed production	1,039	1,119	944	972	998	1,021
Imports from: the Eleven (j)	198	209	238	191	161	171
third countries	47	50	53	47	36	36
Exports to: the Eleven (k)	131	164	120	132	115	125
third countries	28	60	36	35	22	25
Total new supply	1,126	1,153	1,079	1,043	1,058	1,078
Increase in stocks	-1	-7	-22	-23	75	77
Total domestic uses	1,126	1,161	1,101	1,066	984	1,000
Home-fed production as % of total new supply for use in UK	92	97	87	93	94	95
Closing stocks	30	110	88	65	140	217

(a) For comparability with other years, the figures have been adjusted from a 53-week to a 52-week basis.
(b) Average prices at representative markets in England and Wales.
(c) Category change January 1988: formerly 1st quality Hereford/Friesian bull calves.
(d) Category change January 1988: formerly 1st quality yearling steers beef/dairy cross, now consists of Hereford/cross, Charolais/cross, Limousin/cross, Simmental/cross, Belgian blue/cross, other continental/cross, other beef/dairy cross, other beef/beef cross.
(e) Average dressed carcase weight of animals fed and slaughtered in the UK.
(f) Gross indigenous production (GIP) is a measure of animal production commonly used in other EC states and is therefore useful as a means for making international comparisons. It is measured as total slaughterings plus all live exports minus all live imports. GIP differs from home-fed production in that it includes exports and excludes imports of pure bred breeding animals and, for other imported animals, includes only the weight added since their arrival in the country. Both measures include the export weight (dcw equivalent) of animals intended for slaughter abroad.
(g) Average realised return per kg of home-fed production net of marketing expenses. Includes variable premium but not other receipts.
(h) Comprising hill livestock compensatory amounts, suckler cow premium, calf subsidy and beef special premium.
(i) Does not include meat offals or trade in preserved or manufactured meat products. Boneless meat has been converted to bone-in weights.
(j) Includes meat from finished animals imported from the Irish Republic.
(k) Adjusted, as necessary, for unrecorded trade in live animals.

TABLE 5.18 Sheep and lambs; mutton and lamb

Calendar years

	Average of 1980-82	1987(a)	1988	1989	1990	1991 (forecast)
Populations						
Total sheep and lambs ('000 head at June)	32,204	38,756	41,007	42,988	43,799	43,919
of which: ewes	12,538	14,836	15,521	16,205	16,760	17,015
shearlings	2,786	3,287	3,556	3,834	3,650	3,374
lambs under 1 year old	15,638	19,381	20,596	21,564	22,023	22,171
other	1,241	1,252	1,334	1,384	1,365	1,360
Selected market prices						
Store sheep (£ per head):						
1st quality lambs, hoggets and tegs (b)	30.2	37.5	37.5	36.1	33.8	31.5
Finished sheep (p per kg estimated dcw) (c):						
Great Britain	..	196.7	177.0	184.5	174.4	143.4
Northern Ireland	..	213.7	208.0	202.5	170.9	179.3
Marketings, production and returns (excluding clip wool)						
Total home-fed marketings ('000 head):	14,626	16,494	18,143	20,581	21,101	22,329
of which: clean sheep and lambs	13,114	14,842	16,452	18,611	19,000	20,350
ewes and rams	1,512	1,652	1,691	1,970	2,101	1,978
Average dressed carcase weights (dcw) (kgs) (d):						
clean sheep and lambs	18.1	18.0	18.2	18.0	17.8	17.8
ewes and rams	28.5	26.2	25.7	25.8	26.1	26.6
Production ('000 tonnes, dcw):						
Home-fed production	280	311	342	385	393	415
Gross indigenous production (e)	..	309	341	384	391	414
Average realised return (p per kg dcw) (f)	164	207	206	204	181	174
Total realised return (£ million)	459	644	704	784	712	722
Other receipts (£ million) (g)	68	171	198	184	281	353
Value of home-fed production (£ million)	528	815	903	968	993	1,076
Supplies ('000 tonnes, dcw) (h)						
Home-fed production	280	311	342	385	393	415
Imports from: the Eleven (i)	...	...	1	5	7	4
third countries	190	154	136	134	147	124
Exports to: the Eleven (j)	46	88	94	107	98	100
third countries	4	3	2	2	3	3
Total new supply	421	374	383	414	447	440
Increase in stocks	6	-5	-6	-2	6	4
Total domestic uses	415	380	390	416	441	436
Home-fed production as % of total new supply for use in UK	67	83	89	93	88	94
Closing stocks	15	24	18	16	22	26

(a) For comparability with other years, the figures have been adjusted from a 53-week to a 52-week basis.
(b) Average prices at representative markets in England and Wales, excluding prices at autumn hill sheep sales.
(c) Average of weekly market prices as used to determine level of ewe premium.
(d) Average dressed carcase weight of animals fed and slaughtered in the UK.
(e) See footnote (f) to Table 5.17.
(f) Average realised return per kg of home-fed production net of marketing expenses. Includes variable premium but not other receipts.
(g) Comprising hill livestock compensatory amounts and annual ewe premium.
(h) Does not include meat offals or trade in preserved or manufactured meat products. Boneless meat has been converted to bone-in weights.
(i) Includes meat from finished animals imported from the Irish Republic.
(j) Adjusted, as necessary, for unrecorded trade in live animals.

TABLE 5.19 Pigs and pigmeat

Calendar years

	Average of 1980-82	1987(a)	1988	1989	1990	1991 (forecast)
Populations						
Total pigs ('000 head at June)	7,889	7,943	7,982	7,509	7,449	7,659
of which: sows in-pig and other sows for breeding	730	713	703	660	660	676
gilts in-pig	114	107	101	97	109	105
other	7,045	7,123	7,177	6,752	6,680	6,878
Selected market price						
Clean pigs (p per kg deadweight)	85.4	97.6	91.0	113.0	111.9	101.5
Marketings, production and returns						
Total home-fed marketings ('000 head)	14,929	15,656	15,823	14,582	14,294	14,628
of which: clean pigs	14,579	15,308	15,432	14,245	13,962	14,255
sows and boars	351	348	391	337	332	372
Average dressed carcase weights (dcw) (kgs) (b):						
clean pigs	61.7	62.2	62.6	63.0	64.8	65.7
sows and boars	132.7	133.4	134.9	137.4	144.9	145.0
Production ('000 tonnes, dcw):						
Home-fed production	946	998	1,019	944	952	990
Gross indigenous production (c)	..	999	1,020	944	958	993
Average realised return (p per kg dcw) (d)	89	93	86	108	107	96
Value of home-fed production (£ million)	846	930	881	1,023	1,014	952
Supplies of pork ('000 tonnes, dcw) (e) (f)						
Home-fed production	716	778	801	731	749	795
Imports from: the Eleven	29	44	57	90	80	74
third countries	5	5	4	4	1	1
Exports to: the Eleven (g)	31	47	57	59	58	72
third countries	1	4	1	1	1	1
Total new supply	718	777	804	766	771	798
Increase in stocks	-1	1	...	1	2	...
Total domestic uses	719	776	804	765	769	797
Home-fed production as % of total new supply for use in UK	100	100	100	96	97	100
Closing stocks	2	8	8	9	10	11
Supplies of bacon and ham ('000 tonnes, product weight) (e)						
Home-cured production	202	197	199	194	180	174
Imports from: the Eleven (h)	290	259	255	258	259	263
third countries	7	1	1	2	1	1
Exports to: the Eleven	5	5	5	5	5	5
third countries	...	...	...	...	...	...
Total new supply	494	452	450	449	435	433
Increase in stocks	...	2	1	...	-1	1
Total domestic uses	494	450	449	448	436	431
Home-cured production as % of total new supply for use in UK	41	44	44	43	41	40
Closing stocks	2	5	6	6	5	7

(a) For comparability with other years, the figures have been adjusted from a 53-week to a 52-week basis.
(b) Average dressed carcase weight of animals fed and slaughtered in the UK.
(c) See footnote (f) to Table 5.17.
(d) Average realised return per kg of home-fed production net of marketing expenses.
(e) Does not include meat offals or trade in preserved or manufactured meat products.
(f) Boneless meat has been converted to bone-in weights.
(g) Adjusted, as necessary, for unrecorded trade in live animals.
(h) Includes meat from finished animals imported from Irish Republic.

TABLE 5.20 Poultry and poultrymeat

Calendar years

	Average of 1980-82	1987 (a)	1988	1989	1990	1991 (forecast)
Population						
Number ('000 head at June):						
chickens and other table fowls	59,274	70,869	75,437	70,176	73,588	75,228
hens (boiling fowls)	12,138	9,972	8,787	7,569	6,940	7,392
turkeys (b)	7,453	8,840	9,403	9,391	9,596	10,531
ducks	1,389	1,576	1,661	1,909	2,038	2,046
geese	146	181	188	201	179	171
Slaughterings, production and returns						
Slaughterings (millions):						
fowls	421	515	547	496	521	543
turkeys	23	30	31	34	32	33
ducks	7	8	9	11	12	11
geese	...	1	1	1	1	1
Production ('000 tonnes deadweight):						
chickens and other table fowls	572	755	812	751	787	829
boiling fowls (culled hens)	59	53	51	38	39	33
turkeys	123	160	166	177	171	174
ducks	16	18	19	24	25	24
geese	2	3	3	3	3	3
Average producer price (p/kg live weight) for:						
chickens and other table fowls	51.5	55.5	53.8	56.6	59.8	55.9
boiling fowls (culled hens)	20.1	25.6	12.7	19.6	23.7	14.5
turkeys	74.9	90.2	86.5	90.8	90.6	94.7
ducks	77.8	102.3	105.3	116.9	119.8	124.9
geese	156.5	188.2	193.1	193.4	206.6	202.8
Value of output (£ million):						
fowls	409	577	592	577	640	625
turkeys, ducks, geese	136	212	213	247	242	253
Total	545	789	805	824	881	878
Supplies of poultrymeat ('000 tonnes deadweight)						
Production	772	989	1,050	993	1,025	1,063
Imports from: the Eleven	25	82	80	84	135	129
third countries	1	...	...	...	...	...
Exports to : the Eleven	13	40	47	51	47	59
third countries	6	10	14	15	15	21
Total new supply	779	1,021	1,070	1,011	1,097	1,112
Increase in stocks	-4	15	1	-25	8	23
Total domestic uses	783	1,006	1,069	1,036	1,089	1,089
Production as % of total new supply	99	98	98	96	94	98

(a) For comparability with other years, the figures have been adjusted from a 53-week to a 52-week basis.
(b) From 1986 onwards data relates to November and to England and Wales only.

Livestock products

Milk and milk products
(Tables 5.21 and 5.22)

31. Because of continuing oversupply on the dairy market, a 2 per cent reduction in the wholesale and direct sales quotas of each Member State was agreed as part of the price fixing package. Further reductions were authorised to provide quota for allocation to producers entitled to it as a result of further European Court judgements arising from the Non-Marketing of Milk and Dairy Herd Conversion Schemes; Member States were empowered to operate outgoers schemes to allow the total or partial restorations of quota cuts in 1992. Provision was made for compensation to be paid in autumn 1992 to producers whose quota was reduced. A small further uncompensated quota cut was made in the UK to provide quota for allocation to producers of products such as yogurt and ice cream which were brought within the quota system from 1 April 1991. Once adjusted for a high level of butterfat, wholesale milk deliveries exceeded quota and thus resulted in a levy of £9.2 million; £1.4 million was paid in direct sales levy.

32. Producer prices showed only a small increase over those of the previous year due to continued weakness in the dairy commodity markets.

33. Following renewed intervention buying of butter and skimmed milk powder in 1990, after the virtual elimination of stocks in 1989, the Commission took measures to reduce the attractiveness of intervention as an outlet for these products. By means of the tendering arrangements, which were also extended to skimmed milk powder, the accepted bid price was lowered to the point where offers virtually ceased. Buoyant export demand and reduced production following the quota cuts agreed in the prices package led to rising prices for butter and skimmed milk powder and sales out of intervention were re-started in order to ease the market situation.

Eggs *(Table 5.23)*

34. Consumer confidence in eggs continued to recover in 1991 and UK production rose during the year, although it remains below the level of 1988. With the increased supply, prices weakened although they have stabilised and strengthened during the autumn. Imports, which had sharply increased in 1990, fell back during the year but continued above the 1988 level.

Wool *(Table 5.24)*

35. Wool production in the UK is virtually unchanged on last year's levels. The wool price has remained very low in the face of a high level of stocks world-wide and continuing low consumer demand. The guarantee price for the 1991 wool clip was set at 120p per kg, a reduction of 5p per kg from 1990. As a result of the weakness in the wool price, public expenditure on wool is forecast to rise to around £38 million in 1991/92.

36. The Government announced in October 1988 that it intended to end the guarantee arrangements for wool as soon as Parliamentary time permitted. Although that remains the Government's intention, it has not yet been possible to introduce the necessary legislation and it was announced in October 1991 that the wool guarantee would continue to operate for the 1992/93 clip year.

TABLE 5.21 Milk

Million litres (unless otherwise specified)   Calendar years

	Average of 1980-82	1987	1988(a)	1989	1990	1991 (forecast)
Production and output						
Dairy herd (annual average, '000 head) (b)	3,263	3,075	2,955	2,905	2,869	2,786
Average yield per cow (litres per annum)	4,811	4,896	4,974	4,985	5,148	5,143
Production of milk from the dairy herd (c)	15,698	15,058	14,698	14,481	14,771	14,327
Production of milk from the beef herd	35	13	12	11	11	11
less wastage and milk fed to stock	176	261	260	261	261	261
Output for human consumption	15,557	14,810	14,450	14,232	14,521	14,077
Average total return (pence per litre) (d)	13.79	16.43	17.72	19.16	19.29	19.54
Value of output (£ million)	2,145	2,433	2,561	2,727	2,802	2,751
Utilisation of the output for human consumption						
Sales through MMB schemes:						
for liquid consumption	7,097	6,816	6,792	6,793	6,780	6,731
for manufacture	8,080	7,800	7,426	7,146	7,403	6,977
of which: butter (e)	4,098	3,765	3,023	2,806	2,934	2,365
cheese (f)	2,194	2,615	2,894	2,703	2,950	2,921
cream (g)	955	412	449	516	549	632
condensed milk (h)	479	407	411	442	436	432
milk powder - full cream	253	510	546	544	395	485
other	101	91	103	134	139	142
Total sales through schemes (i)	15,184	14,630	14,231	13,951	14,194	13,719
Other utilisations (j)	373	180	219	280	327	358

(a) 366 days.
(b) Dairy herd is defined as cows and heifers in milk plus cows in calf but not in milk, kept mainly for producing milk or rearing calves for the dairy herd.
(c) Excludes suckled milk.
(d) Derived by dividing total value of output by the total quantity of output available for human consumption.
(e) Includes a small quantity of milk utilised to manufacture anhydrous milk fat (AMF).
(f) Includes farmhouse cheese made under milk marketing schemes.
(g) Excludes cream made from residual fat of low fat milk production.
(h) Includes condensed milk used in the production of chocolate crumb plus production of sweetened and unsweetened machine skimmed milk.
(i) Excludes milk sold through schemes but subsequently exported as whole milk for processing outside the UK. The total sales figures for years from 1981 do not equal the sum of sales for liquid consumption and for manufacture as the measurement of liquid sales is now adjusted for waste in transit.
(j) Includes milk consumed in farm households, sales of liquid milk outside schemes, exports, and milk used for farmhouse manufacture of butter, cheese (made outside milk marketing schemes) and cream.

TABLE 5.22 Milk products

'000 tonnes (unless otherwise specified)   Calendar years

	Average of 1980-82	1987	1988	1989	1990	1991 (forecast)
Butter (a)						
Production (b)	186	176	140	130	138	113
Imports from: the Eleven	98	52	54	54	51	47
third countries	102	72	74	64	63	59
Exports to: the Eleven (c)	65	128	115	62	35	27
third countries	7	6	4	4	4	5
Total new supply (c)	314	166	149	182	213	187
Increase in stocks (d)	-4	-103	-138	-27	19	-1
Total domestic uses (c) (d)	318	269	287	209	194	188
Production as % of total new supply for use in UK	59	106	94	71	65	60
Closing stocks (d)	75	220	82	55	74	73
Cheese						
Production (b)	241	266	300	283	316	312
Imports from: the Eleven	113	142	180	160	183	173
third countries	16	18	18	19	19	19
Exports to: the Eleven	10	20	17	20	22	22
third countries	14	16	11	17	19	20
Total new supply	346	390	470	426	477	462
Increase in stocks	2	-10	33	-10	9	-2
Total domestic uses	344	400	437	436	468	464
Production as % of total new supply for use in UK	70	68	64	66	66	67
Closing stocks	109	113	146	136	145	143
Cream - fresh, frozen, sterilized						
Production (b) (e)	78	51	53	60	64	74
Imports from: the Eleven	5	4	3	3	2	3
third countries	...	...	...	...	...	...
Exports to: the Eleven	1	...	1	6	10	22
third countries	...	2	3	3	5	5
Total new supply	82	53	52	54	51	50
Increase in stocks	..	..	..	..	..	..
Total domestic uses	82	53	52	54	51	50
Production as % of total new supply for use in UK	96	96	102	111	125	148
Closing stocks	..	..	..	..	..	..

TABLE 5.22 Milk products (continued)

'000 tonnes (unless otherwise specified)   Calendar years

	Average of 1980-82	1987	1988	1989	1990	1991 (forecast)
Condensed milk (f)						
Production	185	180	183	207	204	201
Imports from: the Eleven	3	11	9	9	11	10
third countries	...	...	...	...	...	...
Exports to: the Eleven	5	6	44	31	8	7
third countries	31	30	25	34	41	34
Total new supply	152	155	123	151	166	170
Increase in stocks	...	2	1	5	-3	...
Total domestic uses	153	153	122	146	169	170
Production as % of total new supply for use in UK	122	116	149	137	123	118
Closing stocks	16	9	10	15	12	12
Milk powder - full cream						
Production	32	94	104	95	70	72
Imports from: the Eleven	2	12	4	4	2	3
third countries	...	1	...	...	...	...
Exports to: the Eleven	4	16	17	27	15	11
third countries	20	36	55	44	35	47
Total new supply	10	55	36	28	22	17
Increase in stocks	1	2	3	...	-4	-1
Total domestic uses	9	53	33	29	25	18
Production as % of total new supply for use in UK	320	171	289	339	319	423
Closing stocks	3	4	7	6	3	2
Skimmed milk powder						
Production	277	194	136	133	166	132
Imports from: the Eleven	8	15	9	16	6	8
third countries	...	1	...	...	...	...
Exports to: the Eleven (c)	68	71	44	45	78	53
third countries	67	38	40	37	26	19
Total new supply (c)	150	101	61	68	68	68
Increase in stocks	40	-22	2	-1	2	-2
Total domestic uses (c)	110	123	59	69	67	70
Production as % of total new supply for use in UK	185	192	223	196	244	194
Closing stocks	77	20	22	21	23	21

(a) From 1980 includes butter other than natural (ie butterfat and oil, dehydrated butter and ghee).
(b) Includes farmhouse manufacture.
(c) These figures include the use of these products for animal feed.
(d) In addition to stocks in public cold stores surveyed by MAFF, closing stocks include all intervention stocks in private cold stores. Total domestic uses does not equate exactly with consumption since changes in unrecorded stocks are not included in the calculation.
(e) Excludes cream made from the residual fat of low fat milk products.
(f) Includes condensed milk used in the production of chocolate crumb plus production of sweetened and unsweetened machine skimmed milk.

TABLE 5.23   Eggs

Calendar years

	Average of 1980-82	1987(b)	1988	1989	1990	1991 (forecast)
Number of layers and output of eggs (a)						
Number of fowls in lay (annual average, millions) (c)	57.62	47.14	46.99	41.09	43.48	44.88
Average yield per layer (number of eggs)	234.7	244.1	243.7	247.8	245.9	246.6
Gross production (c) (million dozen)	1,127	959	954	848	891	922
Output of eggs for human consumption from fowls (excluding waste and eggs used for hatching) (million dozen)	1,061	880	871	764	807	835
Average realised price for eggs from fowls (p per dozen)	46.9	50.4	45.0	51.1	58.3	51.8
Value of output of eggs from fowls (£ million)	497	443	392	391	470	432
Value of output of all eggs (£ million)	500	447	397	397	478	439
Utilisation of UK output for human consumption and other supplies (million dozen) (a)						
Total UK output of hen and duck eggs for human consumption (d)	1,065	884	876	770	813	841
of which: hen eggs sold in shell	1,027	824	803	688	730	757
hen eggs processed	34	56	68	76	77	78
Imports from (e): the Eleven	37	34	38	53	77	57
third countries	2	...	...	...	...	...
Exports to (e): the Eleven	35	17	20	34	13	16
third countries	2	1	1	1	1	...
Total new supply	1,066	900	893	790	877	882
Output as % of total new supply for use in UK	100	98	98	98	93	95

(a) These figures have been revised from those shown in previous editions as a result of changes in survey methodology and grossing-up procedures.
(b) For comparability with other years, the figures have been adjusted from 53-week to a 52-week basis.
(c) Includes breeding flocks.
(d) Includes farmhouse consumption.
(e) Includes shell egg equivalent of whole (dried, frozen and liquid) egg and egg yolk, but excludes albumen.

TABLE 5.24 Wool

Million kg, greasy weight equivalent (unless otherwise specified)   Calendar years

	Average of 1980-82	1987	1988	1989	1990	1991 (forecast)
Skin Wool (valued within output from sheep and lambs)						
Production	12	17	18	20	21	21
Clip Wool (a)						
Production	39	45	49	53	53	53
Producer price for clip (p per kg) (b)	90.0	98.4	97.9	96.8	92.5	85.0
Value of output (£ million)	35	43	48	52	49	45
Supplies						
Total production	51	62	67	73	74	74
Imports from: the Eleven	16	41	31	25	22	26
third countries	103	131	102	83	79	76
Exports to: the Eleven	32	42	34	64	30	33
third countries	18	39	38	38	39	32
Total new supply	120	153	128	80	106	111
Production as % of total new supply for use in UK	43	41	52	91	70	67

(a) Strictly the figures relate to clip years (May/April) but in practice the bulk of the production is within the period May to December.
(b) The price is net of marketing expenses.

6 Agricultural Incomes

Introduction

1. This section provides estimates of agriculture's gross output and input, of its productivity and of the incomes of those engaged, in various ways, in the industry.

Output, input and net product
(Tables 6.1 and 6.2)

2. Table 6.1 begins by drawing together the estimates of the value of output of each of the commodities covered in Section 5. Together with the value of output of various other commodities, and other items (including the value of the physical increase in on-farm stocks), this gives the industry's gross output. Deducting its gross input (expenditure on current inputs adjusted for stock changes) gives its gross product which, after allowing for depreciation of its fixed assets, leads to its net product. This provides the source of remuneration of the various groups providing resources to the industry in the form of financial capital, let land, labour input and managerial skills. The derivation of gross and net product, and of the income measures referred to in this section, is shown diagrammatically in Chart 6.1.

3. Since gross and net product and the resulting income measures are the differences between two large magnitudes (gross output and gross input) it follows that they are very sensitive to changes in either or both of these magnitudes. They are also similarly sensitive to revisions in the estimates of these magnitudes. These points, the revisions now made to previously published figures for earlier years and the 'forecast' nature of the figures for 1991 all need to be borne in mind when considering the forecast changes reported below for 1991. In particular it should be noted that these changes are from higher levels of gross (and net) product and incomes in 1990 than were forecast at this time last year.

4. The value of the industry's gross output is forecast to have changed very little between 1990 and 1991, with a 1.2 per cent rise in volume being largely offset by a 0.9 per cent fall in prices. However the cost of the industry's gross input is forecast to have risen by 3.6 per cent, with a 0.1 per cent reduction in the usage of its inputs being offset by a 3.7 per cent increase in their prices. These changes are reflected in decreases, of 3.1 and 4.6 per cent respectively, in the industry's gross product and net product. Table 6.2 summarises the components of the forecast changes in the industry's gross output and gross input between 1990 and 1991 in both absolute and percentage terms (and separately for the underlying volume and price elements). These changes are also portrayed in Chart 6.2.

Productivity
(Table 6.3)

5. Table 6.3 provides comparisons, over a number of years, of the industry's gross output at constant prices and of two measures of its productivity.

The first, the index of gross agricultural product at constant prices per whole-time man equivalent, has risen by 50.6 per cent over the last decade, including a forecast rise of 5.5 per cent in the last year. However this index does not take account of changes in inputs other than labour. The ratio of the volume of gross output to the volume of productive inputs employed (including labour, usage of capital items and material inputs) provides an alternative measure of productivity. This indicator is forecast to have risen by 21.4 per cent over the last ten years and by 3.4 per cent between 1990 and 1991.

Incomes from farming
(Tables 6.1 and 6.4)

6. The decline of 4.6 per cent, or some £233 million, in the industry's net product was partly offset by a fall of £148 million in the interest paid to the providers of much of the industry's financial capital. The incomes of those engaged in the industry (line 24 in Table 6.1) fell by a much lower amount, £82 million. However since the cost (and earnings) of hired workers rose by £57 million, *total income from farming* (line 26) fell by £138 million (6.0 per cent). This measure provides the basis for the income indicator (3) used by the EC to compare trends in incomes from farming across member states. It reflects the total income from agriculture of the group with an entrepreneurial interest in the industry (farmers and spouses, non-principal partners and directors and their spouses and family workers). In order to derive *farming income*, which covers only farmers and their spouses, it is necessary to attribute earnings to non-principal partners and directors (and their spouses) and family workers and this is done on the basis of the earnings of hired workers. The result is a decrease of £193 million (13.7 per cent) in farming income.

7. Table 6.4 shows movements over the last decade in these income measures (see also Chart 6.3) and in two measures of cash flow which correspond in coverage to total income from farming and farming income. The cash flow measures are intended to reflect more closely the variations as perceived by farmers and farm households. They show very much lower falls between 1990 and 1991 than do the corresponding income measures; ones of only 0.8 per cent for the wider group and of 3.9 per cent for farmers (and their spouses) alone. The table and chart also show the measures expressed in real terms.

Capital formation and stocks
(Tables 6.5 and 6.6)

8. Details of agriculture's investment in productive assets are given in Table 6.5. Total gross fixed capital formation is estimated to have been £1,182 million in 1990, an increase of 5 per cent over 1989. The volume of investment measured at constant 1985 prices showed an increase of 1 per cent between the two years, with increases in the volumes of investment in vehicles and in buildings and works more than offsetting a fall in plant and machinery. Declines are forecast in the stocks of both breeding and all other cattle.

Hired labour
(Table 6.7)

9. Table 6.7 shows the forecast average earnings for whole-time hired men to have increased from £186.10 for a 46.7 hour week in 1990 to £202.20 for a 46.6 hour week in 1991. The earnings figures include estimates of bonuses and of payments made by farmers towards the cost of employees' community charge. The forecasts for 1991 reflect the increase of about 6 per cent in the

statutory minimum wage in the United Kingdom (5.5 per cent in Scotland) and a reduction in hours (except in Northern Ireland). The total cost of hired labour is derived by multiplying the earnings by the numbers employed and taking account of other payments. This has increased by 3.5 per cent to £1,674 million in 1991.

Interest *(Table 6.8)*

10. Details of interest charges payable on farmers' borrowings for current farming purposes and for investment in buildings and works are shown in Table 6.8. Net borrowing for these purposes is forecast to have increased by 1.4 per cent between 1990 and 1991. Over the same period the annual average interest rate paid by farmers on bank advances is forecast to have fallen by 2.8 percentage points to 14.3 per cent, more than offsetting the increase in borrowing. As a result, interest payments are forecast to have fallen by £148 million (14.1 per cent) to £899 million during 1991.

Farm rents
(Table 6.9)

11. Table 6.9 shows that average rents per hectare in England, Wales and Scotland have continued to rise slowly in 1991. For Great Britain as a whole average rent is forecast to have risen by 1.3 per cent between 1990 and 1991. These forecasts are based on provisional results of the annual rent enquiry in England and Wales and continuing field enquiries in Scotland.

TABLE 6.1 Outputs, inputs and net product

£ million Calendar years

	Average of 1980-82	1987	1988	1989	1990	1991 (forecast)
Outputs (a)						
Cereals:						
wheat	926	1,264	1,155	1,368	1,417	1,546
barley	786	692	627	647	599	589
oats	26	29	31	31	34	35
rye, mixed corn and triticale	2	3	3	3	4	4
other receipts (b)	0	2	2	2	1	2
1. Total cereals	1,739	1,990	1,818	2,052	2,055	2,176
Other crops:						
oilseed rape	104	298	241	279	343	324
sugar beet	213	223	239	250	272	269
hops	26	13	13	14	14	18
peas and beans for stockfeed	25	109	142	116	121	110
linseed, fodder and other minor crops (c)	47	79	79	73	89	115
2. Total other crops	415	721	714	732	840	836
Potatoes						
3. Total potatoes	383	486	399	489	508	509
Horticulture						
vegetables (d)	601	959	985	1,028	1,088	1,030
fruit (d)	190	243	255	284	298	299
ornamentals	191	320	389	446	495	503
other (e)	4	9	7	7	6	5
4. Total horticulture	985	1,531	1,636	1,766	1,887	1,837
Livestock:						
finished cattle and calves	1,648	2,069	1,976	2,100	1,973	2,035
finished sheep and lambs	528	815	903	968	993	1,076
finished pigs	846	930	881	1,023	1,014	952
poultry	545	789	805	824	881	878
other livestock (f)	88	120	125	136	146	150
other receipts (g)	8	2	1	0	0	0
5. Total livestock	3,662	4,725	4,691	5,051	5,007	5,091
Livestock products:						
milk	2,145	2,433	2,561	2,727	2,802	2,751
eggs	500	447	397	397	478	439
clip wool	35	43	48	52	49	45
other (h)	18	26	29	42	39	45
6. Total livestock products	2,698	2,950	3,035	3,218	3,368	3,280
Own account capital formation (i):						
breeding livestock	-1	-29	33	40	31	-41
other assets	94	98	115	126	149	149
7. Total own account capital formation	93	69	148	166	180	108
8. Total output (1+2+3+4+5+6+7)	9,974	12,472	12,441	13,475	13,845	13,836
9. Other direct receipts (j)	53	44	96	130	134	143
10. Total receipts (8+9)	10,027	12,516	12,537	13,605	13,979	13,979
Value of physical increase in:						
work-in-progress (k)	-3	-87	-36	-30	-48	-16
output stocks (k)	-26	-44	-8	-63	-8	3
11. Total value of physical increase	-30	-131	-44	-93	-56	-13
12. Gross output (10+11)	9,998	12,384	12,493	13,512	13,923	13,966
Intermediate output (l)						
feed	626	566	575	553	478	586
seed	104	133	131	137	140	138
13. Total	730	699	706	689	618	723
14. Final output (12-13)	9,267	11,686	11,787	12,823	13,305	13,243

TABLE 6.1 Outputs, inputs and net product (continued)

£ million
Calendar years

	Average of 1980-82	1987	1988	1989	1990	1991 (forecast)
Inputs						
Expenditures (net of reclaimed VAT):						
feedingstuffs	2,248	2,614	2,731	2,747	2,848	2,957
seeds	222	282	280	291	298	293
livestock (imported and inter-farm expenses)	167	186	201	185	173	169
fertilisers and lime	727	708	699	744	732	676
pesticides	210	360	440	482	463	444
machinery: repairs	338	544	578	605	646	699
fuel and oil	340	292	258	266	300	321
other	69	108	118	121	123	136
farm maintenance (m)	195	295	300	313	340	354
veterinary expenses and medicines	99	152	157	158	168	177
electricity	113	140	147	155	158	167
miscellaneous expenditures (m)(n)	473	730	772	841	811	917
15. Total expenditure	5,201	6,411	6,681	6,907	7,061	7,310
16. Value of physical increase in input stocks (o)	9	-62	-95	4	15	11
17. Gross input (15-16)	5,192	6,473	6,776	6,903	7,046	7,299
18. Net input (17-13)	4,461	5,774	6,070	6,214	6,428	6,575
19. **Gross product** (12-17) or (14-18)	4,806	5,912	5,717	6,609	6,877	6,667
Depreciation: buildings and works (m)	439	581	611	671	736	772
plant, machinery and vehicles	766	921	975	1,026	1,040	1,026
20. Total depreciation	1,205	1,503	1,586	1,696	1,776	1,799
21. **Net product** (19-20)	3,601	4,409	4,131	4,913	5,101	4,869
22. Interest (p)	526	649	702	946	1,047	899
23. Net rent (m)	87	161	154	142	130	128
24. Income from agriculture of total labour input (21-22-23)	2,988	3,599	3,275	3,825	3,924	3,842
25. Labour: hired (q)	1,122	1,392	1,443	1,496	1,618	1,674
26. Total income from farming (24-25)	1,866	2,207	1,832	2,329	2,306	2,168
27. Labour: family, partners and directors (r)	523	765	803	821	888	944
28. Farming income (s) (26-27)	1,343	1,442	1,029	1,508	1,418	1,224

(a) Output is net of VAT collected on the sale of non-edible products. Figures for total output include subsidies, but not 'Other direct receipts'.
(b) Payment to small-scale cereal producers.
(c) Hay and dried grass, grass and clover seed, root and fodder crop seed, straw, linseed, mustard and other minor crops.
(d) Includes the value of the produce of gardens and allotments.
(e) Seeds, hedgerow fruits and nuts.
(f) Horses, breeding livestock exported, poultry for export, rabbits and game, knacker animals and other minor livestock.
(g) Guidance premium for beef and sheepmeat.
(h) Honey, goats milk, exports of eggs for hatching and minor livestock products.
(i) This comprises the cost of that part of investment in buildings and works which is physically undertaken by the farmer or farm labour and the value of the physical increase in breeding livestock.
(j) This includes the Milk Outgoers Schemes, compensation for milk quota cuts and set-aside payments.
(k) Work in progress is livestock other than breeding livestock. Output stocks comprise cereals, potatoes and some fruits.
(l) Sales included in output but subsequently re-purchased and also included within input.
(m) Landlords' expenses are included within farm maintenance, miscellaneous expenditure and depreciation of buildings and works. Net rent is the rent paid on tenanted land less these landlords' expenses and the benefit value of dwellings on that land.
(n) Including fees, insurance, telephones, and drainage, water and local authority rates (but see reference to farm cottages at (q) below).
(o) Input stocks comprise fertilisers and purchased feed.
(p) Interest charges on loans for current farming purposes and buildings and works less interest on money held on short term deposit.
(q) Including employers' national insurance contributions, perquisites and other payments (including the payment by farmers of rates on farm cottages occupied by farm workers and of their community charge).
(r) The estimate in respect of family workers, non-principal partners and directors (and their spouses) is calculated on the basis of the earnings of hired labour.
(s) The return to farmers (and their spouses) for their labour, management skills and own capital invested after providing for depreciation.

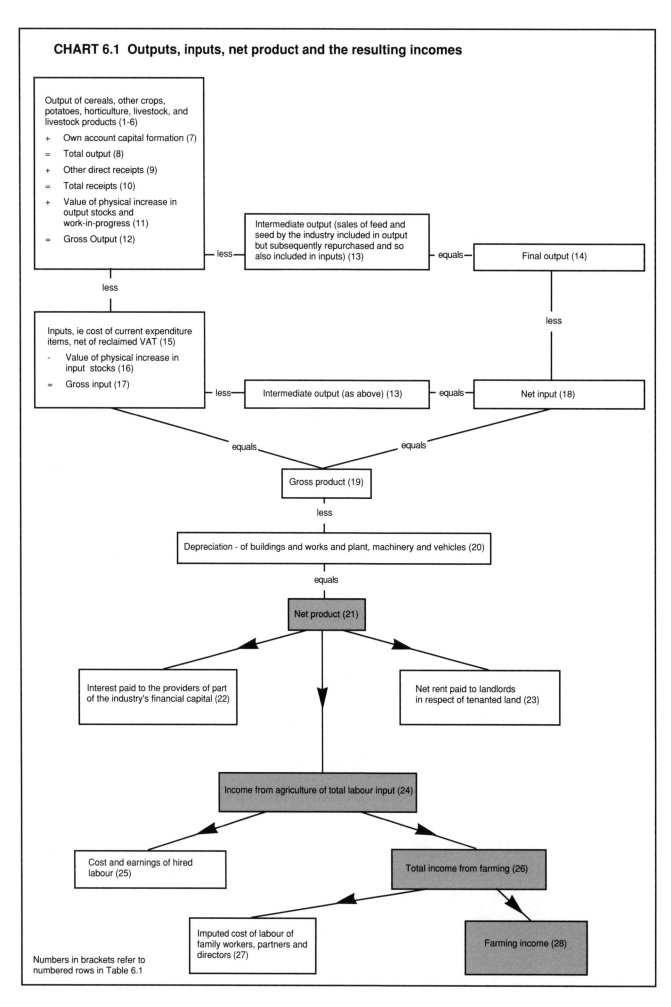

TABLE 6.2 Changes in outputs and inputs

	Change between 1990 and 1991 (forecast)			
	Total change		Percentage change in	
	£ million	Percentage	Price	Quantity
Outputs				
Cereals	121	5.9	4.5	1.3
Other crops	-4	-0.5	-5.1	4.9
Potatoes	2	0.3	-0.4	0.7
Horticulture	-51	-2.7	-1.5	-1.2
Livestock	84	1.7	-1.8	3.6
Livestock products	-88	-2.6	-0.4	-2.2
Other items (a)	-21	..	..	..
Gross output	43	0.3	-0.9	1.2
Inputs				
Feedingstuffs	109	3.8	1.4	2.4
Seeds	-5	-1.8	3.3	-4.9
Livestock	-4	-2.0	5.3	-6.9
Fertilisers and lime	-57	-7.7	-2.1	-5.8
Pesticides	-19	-4.1	11.0	-13.6
Machinery (total current expenses)	87	8.1	7.2	0.9
Farm maintenance	13	3.9	4.2	-0.2
Miscellaneous (inc vets and electricity)	124	10.9	5.4	5.2
Other items (b)	4	..	..	..
Gross input	253	3.6	3.7	-0.1
Gross product	-210	-3.1	-5.5	2.6

(a) Covers own account capital formation, other direct receipts and the value of the physical increase in output stocks and non-breeding livestock work-in-progress.
(b) Covers value of the physical usage of feed and fertiliser stocks.

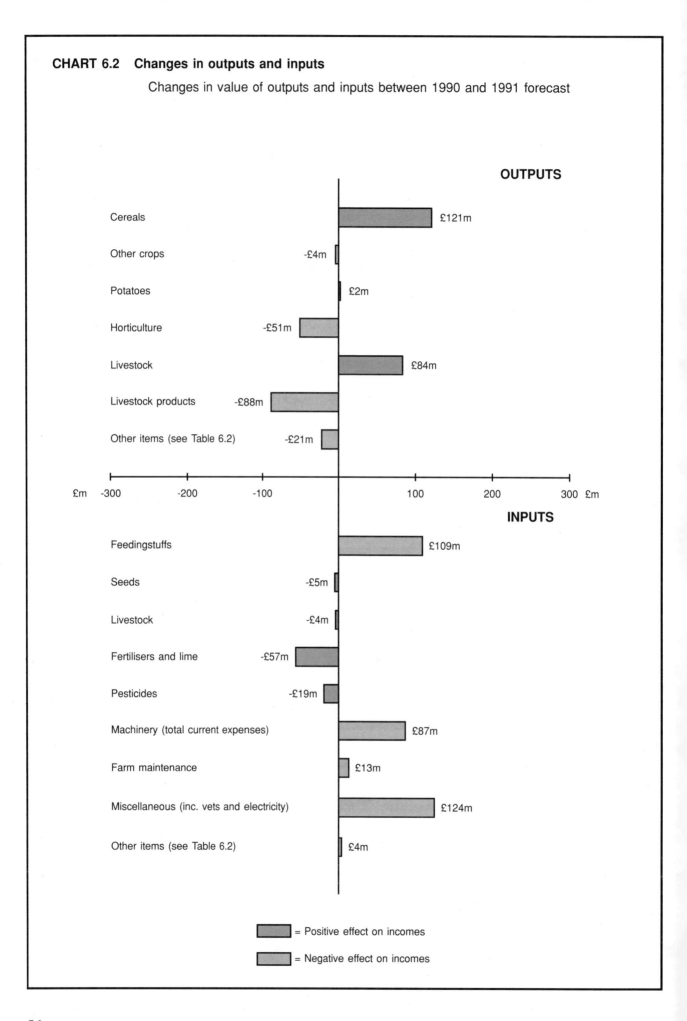

TABLE 6.3 Output volume and productivity

Calendar years: 1985 = 100

Year	Gross output (as defined in Table 6.1) at constant 1985 prices	Gross product (as defined in Table 6.1) at constant 1985 prices per whole-time man equivalent (a)	Gross output per unit of all inputs (including fixed capital and labour), at constant prices
1980	91.9	81.5	91.2
1981	90.9	83.4	92.6
1982	97.0	90.6	96.1
1983	95.7	85.1	92.7
1984	103.0	105.0	102.6
1985	100.0	100.0	100.0
1986	100.5	102.4	101.0
1987	99.4	101.1	100.4
1988	99.0	101.4	100.7
1989	100.5	110.8	104.3
1990	102.1	119.0	108.7
1991 (forecast)	103.3	125.6	112.4

(a) The total numbers of whole-time man-equivalents engaged in agriculture is estimated for this series from the total number of full-time, part-time and casual workers, salaried managers, farmers, partners and directors (and their spouses) returned in the annual June Censuses weighted by their estimated average annual hours worked.

TABLE 6.4 Summary measures from the aggregate agricultural account

Calendar years

Year	Net product (as defined in table 6.1)	Income from farming		Cash flow from farming	
		Total income from farming (of farmers, non-principal partners and directors and their spouses and family workers)	Farming income (of farmers and spouses)	of farmers, non-principal partners and directors and their spouses and family workers	of farmers and spouses
£ million					
1980	3,093	1,473	1,003	1,776	1,306
1981	3,568	1,845	1,322	2,219	1,696
1982	4,142	2,280	1,704	2,449	1,873
1983	3,854	1,923	1,293	2,080	1,451
1984	4,795	2,739	2,093	2,853	2,207
1985	4,009	1,695	995	2,232	1,533
1986	4,392	2,154	1,412	2,578	1,837
1987	4,409	2,207	1,442	2,977	2,212
1988	4,131	1,832	1,029	2,391	1,589
1989	4,913	2,329	1,508	2,943	2,122
1990	5,101	2,306	1,418	2,879	1,991
1991 (forecast)	4,869	2,168	1,224	2,856	1,912
Indices in real terms (deflated by RPI: 1985=100)					
1980	109.1	123.0	142.5	112.6	120.5
1981	112.5	137.6	167.9	125.7	139.9
1982	120.3	156.6	199.3	127.8	142.3
1983	107.0	126.3	144.6	103.8	105.4
1984	126.8	171.4	223.0	135.6	152.7
1985	100.0	100.0	100.0	100.0	100.0
1986	106.0	122.9	137.2	111.7	115.9
1987	102.1	120.9	134.5	123.9	134.0
1988	91.2	95.7	91.5	94.8	91.7
1989	100.6	112.8	124.4	108.3	113.7
1990	95.5	102.1	106.8	96.8	97.4
1991 (forecast)	86.2	90.8	87.3	90.8	88.6

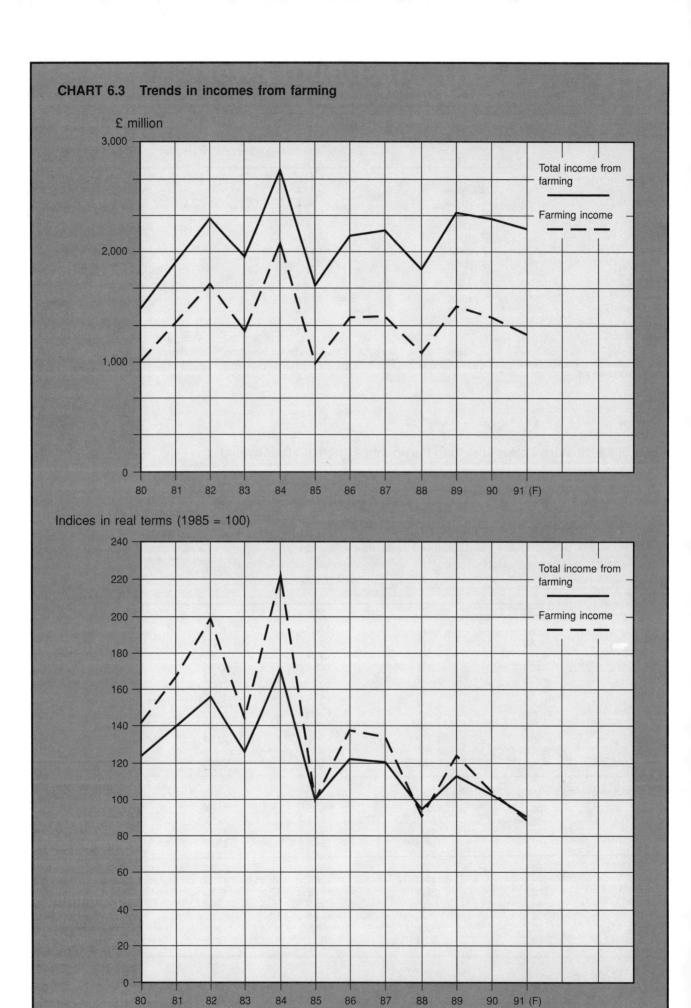

TABLE 6.5 Gross capital formation

£ million   Calendar years

	Average of 1980-82	1987	1988	1989	1990	1991 (forecast)
Gross capital formation at current prices						
Gross fixed capital formation:						
buildings and works	564	396	447	473	540	..
plant and machinery	423	447	538	489	431	..
vehicles	99	126	140	161	211	..
Total	1,086	968	1,125	1,123	1,182	..
Breeding livestock capital formation:						
cattle	-10	-46	23	17	20	-47
sheep	9	18	22	22	8	5
pigs	-	-1	-12	2	3	-
Total	-1	-29	33	40	31	-41
Gross capital formation	1,084	939	1,158	1,163	1,213	..
Gross capital formation at constant 1985 prices						
Gross fixed capital formation:						
buildings and works	586	357	380	367	414	..
plant and machinery	500	434	474	395	334	..
vehicles	125	109	112	120	145	..
Total	1,212	899	965	882	893	..
Breeding livestock capital formation	-3	-29	32	36	26	-34
Gross capital formation	1,209	871	998	918	919	..

TABLE 6.6 Stocks and work in progress

£ million
Calendar years

	Average of 1980-82	1987	1988	1989	1990	1991 (forecast)
Increase in book value of stocks and work in progress	238	-118	26	24	113	49
Stock appreciation	259	76	165	113	153	51
Value of physical increase in stocks and work in progress:						
at current prices	-21	-193	-139	-89	-40	-2
at constant (1985) prices	-25	-203	-146	-75	-33	...
Details at current prices:						
Output stocks:						
wheat	2	-30	-12	1	9	4
barley	-21	-10	-3	-47	-10	1
oats	-1	...	1	3	...	...
potatoes	-4	5	3	-24	1	-13
fruit	-2	-9	2	4	-8	11
Total	-26	-44	-8	-63	-8	3
Work in progress:						
cattle	-6	-110	-34	-4	-50	-25
sheep	...	21	8	-6	3	-3
pigs	4	-1	-8	-10	-1	3
poultry	-1	3	-2	-10	...	10
Total	-3	-87	-36	-30	-48	-16
Input stocks:						
feedingstuffs	3	-27	-23	-10	11	11
fertilisers	6	-35	-72	14	4	...
Total	9	-62	-95	4	15	11

TABLE 6.7 Costs and earnings of hired labour

Calendar years

	Average of 1980-82	1987	1988	1989	1990	1991 (forecast)
Hired labour costs (£ million)						
Wages and salaries (a)	997	1,283	1,330	1,376	1,490	1,543
Insurance payments	123	102	106	112	120	125
Other payments (b)	3	7	7	8	7	7
Total	1,122	1,392	1,443	1,496	1,618	1,674
Hours and earnings of regular whole-time male workers, 20 years and over						
Hours per week (c)	46.0	46.5	46.6	46.6	46.7	46.6
Earnings per week (£) (d)	95.9	148.0	155.1	167.4	186.1	202.2
Index of earnings in real terms, (deflated by the RPI, 1985 = 100)	91.1	102.4	102.3	102.5	104.1	106.9

(a) Includes perquisites.
(b) Includes redundancy payments, Workers' Pension Scheme and Youth Training Scheme.
(c) All hours worked and statutory holidays.
(d) Includes pay for statutory holidays, employers' contribution to employees' community charge and payments in kind for houses, board and lodging and milk which are valued at rates set down by the Agricultural Wages Boards.

TABLE 6.8 Interest

£ million (unless otherwise specified)   Calendar years

	Average of 1980-82	1987	1988	1989	1990	1991 (forecast)
Interest rates						
average bank base lending rate in the UK (percentage)	13.8	9.7	10.1	13.9	14.8	11.7
average rate of interest on bank advances to agriculture (percentage)	16.2	12.1	12.5	16.2	17.1	14.3
Interest charges (for current farming purposes and building and works) on:						
bank advances	481	593	636	881	978	830
AMC loans	5	7	7	9	16	20
instalment credit	21	29	34	48	57	51
leased assets	25	42	54	55	57	50
other credit (a)	3	4	4	5	5	5
less interest on deposits (b)	9	24	31	51	66	56
Total	526	649	702	946	1047	899

(a) This includes interest paid on loans from the Agricultural Credit Corporation, from the Department of Agriculture for Northern Ireland and from private sources.
(b) Interest earned on money held on short term deposit.

TABLE 6.9 Farm rents

Average farm rents per hectare

Index numbers (1985=100)   Calendar years

	Average of 1980-82	1987	1988	1989	1990	1991 (provisional)
England	66.1	109.7	110.8	111.2	112.5	113.8
Wales	60.6	103.7	108.5	112.1	115.5	119.2
Scotland	64.1	109.9	112.3	113.3	118.1	120.3
Great Britain (a)	65.5	109.4	110.7	111.4	113.0	114.5

(a) Virtually all agricultural land in Northern Ireland is owner-occupied.

7 Land prices and balance sheets

Agricultural land prices
(Table 7.1)

1. This section reports on developments in average land prices and the aggregate balance sheet for agriculture. Land prices are obtained from Inland Revenue statistics of average sale prices. Only a very small proportion of the total area of farmland in the UK is sold in a particular year. The average prices recorded by the Inland Revenue can therefore be subject to substantial variation from year to year and, in the case of the unweighted averages shown in Table 7.1, may vary with size and type of lot sold in the year concerned.

2. In 1990, however, the unweighted average price per hectare of land sold with vacant possession recorded in the Inland Revenue series changed only slightly in England, Scotland and Northern Ireland. In Wales it fell by about 20 per cent from the peak reached in 1989. The average price realised for sales in the tenanted sector changed very little over the period.

Balance sheet
(Table 7.2)

3. Estimates of the aggregate balance sheet at current prices shown in Table 7.2 indicate that in 1990 the total value of assets (net of depreciation) fell back by about 10 per cent from the peak level of 1989. This was largely due to the fall in the value of land and buildings which, for the purpose of estimating the value of assets, is based on the sale prices weighted by the proportion of all land in the UK in each farm size group and region. The total value of liabilities rose by 3.5 per cent in 1990 and the industry's net worth fell by around 13 per cent. In real terms net worth fell between 1989 and 1990 by about 20 per cent.

TABLE 7.1 Agricultural land prices

£ per hectare   Calendar years

	Average of 1980-82	1986	1987	1988	1989	1990 (b)
England (a)						
With vacant possession	3,519	3,397	3,516	4,220	4,746	4,683
Tenanted	2,425	2,070	2,304	3,061	2.135	2,131
Wales (a)						
With vacant possession	2,255	2,480	1,912	3,693	3,909	3,145
Scotland (a)						
With vacant possession	1,818	1,561	1,456	1,458	1,412	1,440
Tenanted	1,357	761	762	716	756	740
Northern Ireland (a)						
With vacant possession	2,936	3,128	3,204	2,855	3,359	3,464

(a) These series, based on Inland Revenue data, exclude land sold for non-agricultural purposes. In Great Britain sales of less than 5 hectares and in Northern Ireland of less than 2 hectares are also excluded. In Scotland the series refers to sales of equipped farms only and excludes sales of whole estates and inter-family sales. There is a delay between the date on which a sale is agreed and the date on which it is included in the analysis. The delay is thought to average about 9 months for England and Wales and about 3 months for Northern Ireland. The average prices shown in the table for each calendar year relate to sales included in the series for these countries in the years ending in the following September and March respectively. In the case of Scotland the problem is overcome by further analysis of information by date of sale. The data for Scotland and Northern Ireland are subject to retrospective revision. Reliable prices for tenanted land in Wales are not available due to insufficient sales and virtually all land in Northern Ireland is owner-occupied.
(b) For Scotland and Northern Ireland figures for the most recent years are based on sales notified up to June and March 1991 respectively.

TABLE 7.2 Aggregate balance sheets for agriculture

£ million   As at December each year

	Average of 1980-82	1986	1987	1988	1989	1990 (provisional)
At current prices						
Assets						
fixed (a): land and buildings	35,250	33,600	34,350	40,000	46,550	40,500
plant, machinery and vehicles	4,200	4,650	4,700	5,000	5,250	5,100
breeding livestock	3,500	3,050	3,800	3,650	4,050	3,800
current	5,800	7,100	6,950	7,200	7,650	7,500
Total	48,750	48,450	49,800	55,800	63,500	56,900
Liabilities						
bank loans and overdrafts	3,500	5,600	5,600	6,000	6,400	6,550
other	1,950	2,750	3,000	3,200	3,600	3,800
Total	5,450	8,400	8,600	9,200	10,000	10,350
Net worth	43,300	40,050	41,200	46,600	53,500	46,550
Indicators in real terms (deflated by RPI, Dec 1985=100)						
Total assets	121	93	92	97	102	84
Total liabilities	82	98	97	98	98	93
Net worth	129	92	91	96	103	82

(a) The valuations of land, buildings and breeding livestock are at average market prices; those of plant, machinery and vehicles are at replacement cost, net of depreciation.

8 Farm business data

Introduction

1. Information on incomes, assets and liabilities of full-time farm businesses in the United Kingdom is provided by the annual Farm Business Survey which is conducted by universities and agricultural colleges in Great Britain and by the Department of Agriculture for Northern Ireland. Summary results of these surveys (weighted according to the distribution of holdings by type, size and tenure recorded in the June Census) are presented and described in this section, together with forecasts of farm incomes for the 1991/92 year.

2. It should be noted that the accounting practices and concepts adopted in the Farm Business Survey differ in a number of respects from those employed in compiling the aggregate account as reported in Section 6. Income measures deriving from the two sources are not, therefore, directly comparable. All data in this section are averages per farm.

Farm incomes
(Tables 8.1 - 8.3)

3. Movements in *net farm income* over recent years, for each country and for the main farm types, are shown in Table 8.1. This income measure is a long-standing indicator of the economic performance of farm businesses and, in order to achieve comparability among farms of different types of tenure, it is based on the assumption that all land is tenanted. It represents the return to the farmer and spouse for their manual and managerial labour and on the tenant-type assets of the business such as livestock, machinery and permanent crops (but not land and buildings).

4. Figures in Table 8.1 for 1990/91 (an accounting year ending on average in February 1991) show that incomes on dairy farms in each country declined and, for the UK as a whole, fell to around their 1986/87 levels in real terms. For hill and upland (LFA) livestock farms and lowland livestock farms, there were significant reductions in incomes for most countries. Incomes on cereal and, in nominal terms, other cropping farms showed an increase for the second year running but pig and poultry farms experienced a decline in incomes following a substantial rise in the previous year.

5. Forecasts of net farm income for 1991/92 are based on information from a wide variety of sources and assume normal weather patterns between November 1991 and February 1992. The income forecasts should be regarded only as broad indicators of the overall effects on income of expected developments in output values and input costs.

6. Dairy farm incomes are expected to decline further in 1991/92 due mainly to cuts in milk quota. Following large increases in sheep subsidies in

the UK (sheep annual premium and the new 1990/91 rates for HLCAs), incomes on hill and upland (LFA) livestock farms are expected to recover and, in all countries, to be broadly similar to the levels of 1989/90. Incomes on lowland livestock farms in England are forecast to increase in 1991/92 due to a recovery in cattle prices. In Scotland, however, incomes on these farms are forecast to fall. This is mainly due to the relatively poor performance of cropping enterprises. Cereal yields in Scotland have not matched the levels of 1990/91 and incomes on cropping farms look set to decline. In contrast, incomes on cereal and other cropping farms in England and Wales are forecast to increase in 1991/92 mainly due to increases in the yield and price of wheat and the price of potatoes. As a result of low finished pig and poultry prices, incomes on pig and poultry farms are set to decline markedly.

7. Information on the actual levels of net farm income in 1989/90 and 1990/91 is shown in Table 8.2 according to farm type, country and business size. Farm business size is measured in financial terms, based on standard gross margins per hectare of crops and per head of livestock. For each size group, pig and poultry farms showed the highest level of income in both years and dairy farms showed larger incomes on average than cropping farms and LFA and lowland livestock farms. Income levels fell between 1989/90 and 1990/91 for each size and type of farm in each country except for small and medium LFA livestock farms in Scotland and medium and large cropping farms in England and Scotland.

8. *Occupier's net income* is an alternative measure of farm performance which represents the return to the farmer and spouse for their manual and managerial labour and on all assets invested in the farm business, including land and buildings. It takes account of the actual expenditure associated with owning or renting land. By measuring farm income after the payment of rent and interest charges, it may reflect more realistically changes in income as perceived by farmers. Table 8.3 shows levels of occupier's net income by farm type, country and tenure. Movements in occupier's net income were generally similar to those in net farm income between 1989/90 and 1990/91.

Assets and liabilities
(Table 8.4)

9. Table 8.4 provides information on the assets, liabilities and net worth of farm businesses, according to country and type of tenure in 1989/90 and 1990/91. Continuing the trend of recent years, total assets per farm increased on average, in nominal terms, across all countries and tenure types, apart from tenanted farms in Wales which showed a decline. There was a general increase in the value of fixed assets, again with the exception of tenanted farms in Wales. Average net worth across all tenure types and countries showed an increase apart from tenanted farms in England and Wales and mixed tenure farms in Wales. Liabilities increased for each tenure type and country except for tenanted farms in Scotland.

10. External liabilities, expressed as a percentage of total assets, give an indicator of indebtedness. In England this ratio (based on closing valuations) increased slightly between 1989/90 and 1990/91 to 29 per cent on tenanted farms. It was broadly unchanged at 11 per cent on owner-occupied farms and

14 per cent on farms with mixed tenure. Figures for Wales showed small increases in the ratios to 19 per cent for tenanted farms, 8 per cent for owner-occupied farms and 13 per cent on mixed tenure farms. In Scotland ratios were around 20 per cent for wholly or mainly tenanted farms and 17 per cent for wholly or mainly owner-occupied farms. Nearly all farms in Northern Ireland are owner-occupied and for these farms average liabilities increased slightly to around 6 per cent of total assets.

TABLE 8.1 Net farm income by country and farm type

Indices of average net farm income per farm (1982/83=100) Accounting years ending on average in February

Country and farm type	1986/87	1987/88	1988/89	1989/90	1990/91	1991/92 (forecast)
At current prices						
England:						
Dairy	85	103	132	122	97	90
LFA livestock	73	100	123	82	59	85
Lowland livestock	12	29	25	11	5	10
Cereals	53	9	7	22	30	35
Other cropping	107	45	43	114	117	120
Pigs and poultry	111	94	56	228	204	90
Wales:						
Dairy	92	123	152	151	119	115
LFA livestock	119	128	180	113	84	120
Scotland:						
Dairy	49	119	170	199	157	130
LFA livestock	44	104	117	112	111	110
Lowland livestock	24	52	62	54	66	40
Cropping	88	58	30	87	93	45
Northern Ireland:						
Dairy	51	109	128	119	65	65
LFA livestock	34	125	107	57	50	60
United Kingdom:						
Dairy	80	108	136	129	100	90
LFA livestock	71	109	133	96	79	95
Lowland livestock	16	34	31	14	10	15
Cereals	55	14	10	31	41	45
Other cropping	113	49	45	119	121	125
Pigs and poultry	111	94	56	228	204	90
In real terms (deflated by the RPI)						
United Kingdom:						
Dairy	66	86	103	91	64	55
LFA livestock	59	87	100	67	51	60
Lowland livestock	14	27	24	10	6	10
Cereals	46	11	7	22	26	25
Other cropping	94	39	34	83	78	75
Pigs and poultry	93	75	43	160	130	55

TABLE 8.2 Net farm income by farm type, country and size

With comparative data on average farm area and number of holdings

Accounting years ending on average in February

Farm type and country	Small		Medium		Large		All size groups		Average farm area including rough grazing (hectares per farm) 1990/91			Number of holdings at June 1990		
	1989/90	1990/91	1989/90	1990/91	1989/90	1990/91	1989/90	1990/91	Small	Medium	Large	Small	Medium	Large
Dairy:														
England	9.7	8.2	21.2	16.1	43.8	36.5	24.3	19.4	31	64	140	5,718	11,450	5,323
Wales	11.5	10.2	23.2	19.0	59.9	42.1	24.1	19.0	28	64	139	1,926	2,429	634
Scotland	..	..	20.0	15.0	39.6	32.4	27.5	21.9	..	73	148	153	1,222	1,084
N. Ireland	9.0	4.8	26.3	14.5	..	..	15.2	8.3	41	73	..	3,618	2,028	250
LFA livestock:														
England	3.5	2.0	11.8	9.3	17.7	12.5	8.2	5.9	83	261	559	4,391	3,142	1,005
Wales	4.1	2.6	13.2	9.6	43.5	37.4	10.1	7.6	87	278	603	5,167	3,346	709
Scotland	5.9	6.2	11.9	12.2	25.2	21.6	10.1	9.9	273	423	981	4,828	3,264	919
N. Ireland	2.8	2.6	9.5	6.1	..	..	3.3	2.9	90	204	..	3,914	310	29
Lowland livestock:														
England	2.7	1.5	4.7	0.4	10.6	9.6	3.7	1.8	48	99	187	10,849	3,876	1,196
Scotland	..	..	8.3	8.1	..	..	8.1	9.8	..	87	..	391	491	231
Cropping:														
England	4.3	3.2	6.0	7.2	31.7	33.8	17.6	18.9	42	83	252	9,496	11,079	14,030
Scotland	..	..	6.6	7.1	29.8	30.6	14.7	15.6	..	82	194	780	1,550	1,500
Pigs and poultry:														
England	15.7	14.8	33.7	22.8	85.2	84.1	42.1	37.6	9	22	72	1,867	1,618	1,416

(a) Figures are not shown separately where the sample contains fewer than 20 farms.

TABLE 8.3 Occupier's net income by farm type, country and tenure

With comparative data on average farm area and number of holdings

Accounting years ending on average in February

Farm type and country	Farm Business Survey data (a)								Average farm area including rough grazing (hectares per farm) 1990/91			Census data		
	Occupier's net income (£'000 per farm)											Number of holdings at June 1990		
	Owner-occupied		Tenanted		Mixed tenure		All types of tenure		Owner-occupied	Tenanted	Mixed tenure	Owner-occupied	Tenanted	Mixed tenure
	1989/90	1990/91	1989/90	1990/91	1989/90	1990/91	1989/90	1990/91						
Dairy:														
England	21.5	15.1	17.3	11.6	23.5	18.2	21.1	15.3	70	67	90	9,919	5,440	7,132
Wales	19.6	14.9	..	..	28.5	17.6	21.6	15.7	57	..	81	2,893	849	1,247
Scotland (b)	26.7	21.2	..	..	..	..	26.4	20.6	98	..	..	1,786	673	..
N. Ireland (c)	12.3	4.9	..	..	..	..	..	..	50	..	..	5,896	..	..
LFA livestock:														
England	3.4	0.7	6.4	4.5	10.2	5.2	6.0	2.8	122	291	266	4,012	2,061	2,465
Wales	9.8	6.9	..	..	7.6	4.5	8.8	5.8	173	..	234	5,775	1,239	2,208
Scotland (b)	8.3	7.2	8.7	9.5	..	..	8.4	8.1	274	523	..	5,528	3,483	..
N. Ireland (c)	1.2	0.5	..	..	..	..	..	..	84	..	..	4,253	..	..
Lowland livestock:														
England	3.7	2.0	1.9	-3.2	3.7	2.8	3.5	1.6	65	84	79	9,316	2,735	3,870
Scotland (b)	..	..	..	..	..	..	8.4	12.3	..	..	..	757	356	..
Cropping:														
England	12.8	14.7	11.6	11.3	21.3	22.5	15.5	16.7	145	143	172	14,338	8,202	12,065
Scotland (b)	10.2	13.9	17.3	17.5	..	..	12.9	15.3	116	120	..	2,392	1,438	..
Pigs and poultry:														
England	31.4	29.2	..	..	49.9	41.4	32.9	29.8	26	..	60	3,535	500	866

(a) Figures are not shown separately where the sample contains fewer than 20 farms.
(b) All survey farms in Scotland are classified according to the main tenure category of land on the holding.
(c) Practically all farms in Northern Ireland are owner-occupied.

TABLE 8.4 Assets and liabilities of farm businesses by country and tenure

Closing valuations

£'000 per farm

Accounting years ending on average in February

		Owner-occupied		Tenanted		Mixed tenure		All types of tenure	
		1989/90	1990/91	1989/90	1990/91	1989/90	1990/91	1989/90	1990/91
England	Total assets	546.3	554.1	148.4	149.3	536.3	551.5	463.8	472.4
	of which:fixed assets	489.9	497.4	93.4	95.1	463.8	478.2	402.8	411.3
	current assets	56.4	56.7	55.0	54.2	72.5	73.3	61.0	61.2
	Total external liabilities	61.0	63.5	40.2	42.7	75.6	78.9	61.2	64.0
	of which:long and medium term loans	24.5	26.0	5.7	5.4	27.4	28.1	21.6	22.5
	short-term loans	36.5	37.4	34.4	37.3	48.2	50.8	39.6	41.4
	Net worth	485.3	490.6	108.2	106.6	460.8	472.6	402.6	408.5
	Occupier's net income 1990/91 year		12.4		9.5		17.4		13.3
Wales	Total assets	399.1	409.8	95.7	91.5	355.9	360.7	350.8	358.3
	of which:fixed assets	377.0	387.7	78.3	75.4	322.6	329.6	326.8	334.9
	current assets	22.1	22.2	17.4	16.1	33.3	31.1	24.0	23.4
	Total external liabilities	27.4	32.0	16.2	17.7	41.2	48.0	29.1	33.8
	of which:long and medium term loans	11.3	12.3	1.7	1.8	16.7	18.7	11.3	12.4
	short-term loans	16.1	19.7	14.5	15.9	24.5	29.3	17.8	21.4
	Net worth	371.7	377.8	79.5	73.8	314.7	312.7	321.7	324.5
	Occupier's net income 1990/91 year		8.6		8.1		8.1		8.5
Scotland (a)	Total assets	323.5	331.7	134.0	137.2	:	:	254.9	261.2
	of which:fixed assets	269.6	274.4	74.3	75.1	:	:	198.8	202.1
	current assets	53.9	57.3	59.8	62.1	:	:	56.1	59.0
	Total external liabilities	54.7	56.9	29.2	28.0	:	:	45.5	46.4
	of which:long and medium term loans	10.0	9.3	4.8	4.4	:	:	8.1	7.5
	short-term loans	44.7	47.5	24.5	23.6	:	:	37.4	38.9
	Net worth	268.8	274.8	104.8	109.2	:	:	209.3	214.8
	Occupier's net income 1990/91 year		11.1		13.3	:	:		11.9
Northern Ireland (b)	Total assets	238.7	242.9	:	:	:	:	:	:
	of which:fixed assets	210.0	214.8	:	:	:	:	:	:
	current assets	28.7	28.1	:	:	:	:	:	:
	Total external liabilities	12.0	14.1	:	:	:	:	:	:
	of which:long and medium term loans	5.8	6.1	:	:	:	:	:	:
	short-term loans	6.2	8.0	:	:	:	:	:	:
	Net worth	226.7	228.8	:	:	:	:	:	:
	Occupier's net income 1990/91 year		2.7	:	:	:	:	:	:

(a) All survey farms in Scotland are classified according to the main tenure category of land on the holding.
(b) Practically all farms in Northern Ireland are owner-occupied.

9 Public expenditure on agriculture

Introduction

1. Table 9.1 shows public expenditure under the CAP and on national grants and subsidies, while Table 9.2 provides more detailed information on the costs of market regulation under the CAP. The tables exclude other expenditure which may benefit farmers (eg expenditure on animal health or on research, advice and education). They do, however, include some expenditure which benefits consumer and trade interests rather than producers directly. The figures for the financial years up to and including 1990/91 represent actual expenditure recorded in the Appropriation Accounts. The figures for 1991/92 are the latest estimates of expenditure.

Public expenditure
(Tables 9.1 and 9.2)

2. Expenditure in the United Kingdom on market regulation under the CAP is estimated to be some £1,736 million in 1991/92 compared to about £1,623 million in 1990/91. Chart 9.1 illustrates the breakdown for each commodity sector. This expenditure includes export refunds and monetary compensatory amounts; the net cost of buying, storing and selling commodities taken into intervention; and a variety of grants and subsidies such as the

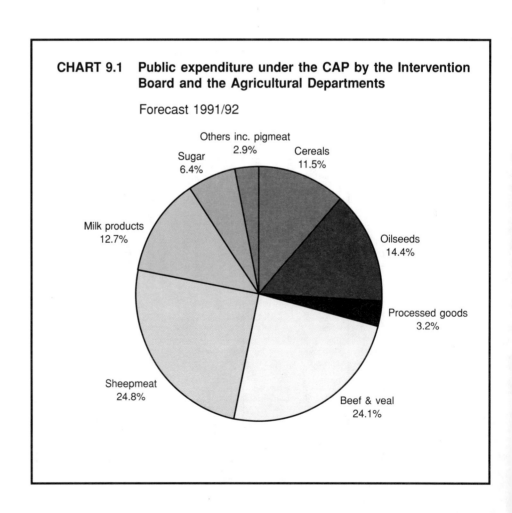

CHART 9.1 Public expenditure under the CAP by the Intervention Board and the Agricultural Departments

Forecast 1991/92

- Others inc. pigmeat 2.9%
- Cereals 11.5%
- Sugar 6.4%
- Milk products 12.7%
- Oilseeds 14.4%
- Processed goods 3.2%
- Sheepmeat 24.8%
- Beef & veal 24.1%

sheep variable premium scheme (which ends in January 1992), the beef special premium and suckler cow premium schemes, the annual premium on ewes, the oilseeds crushing subsidy and payments to producers giving up milk production. Receipts from the milk co-responsibility and supplementary levies and the cereals co-responsibility levy are netted off expenditure on those commodities.

3. The major changes from 1990/91 to 1991/92 are increased expenditure on the sheep variable premium scheme, the annual premium on ewes, cereals export refunds and the oilseeds crushing subsidy. Lower overall expenditure on intervention activity is forecast with more sales of all commodities, lower purchases of beef and dairy products but increased purchases of cereals; repayment of additional cereals co-responsibility levy has ceased and receipts of milk supplementary levy are expected.

4. Other expenditure on agricultural support in the United Kingdom is estimated to be £350 million in 1991/92 compared with £287 million in 1990/91. This expenditure includes capital grants, assistance for agriculture in special areas and price guarantees. Increased expenditure in 1991/92 is mainly due to further substantial sums being needed for the wool guarantee, as a result of severely depressed world prices; increased payments for Hill Livestock Compensatory Allowances and additional uptake under the Set-Aside Scheme.

Intervention stocks
(Table 9.3)

5. Table 9.3 shows the level of opening and closing stocks and purchases into, and sales out of, intervention in the United Kingdom for the years 1987 to 1991/92. This indicates how stocks of cereals and dairy products, after increasing in 1990/91, are forecast to decrease in 1991/92; a continued increase in beef stocks is however expected.

TABLE 9.1 Public expenditure under the CAP and on national grants and subsidies

£ million   April/March (financial years)

	1987/88	1988/89	1989/90	1990/91	1991/92 (forecast)
I Market regulation under the CAP					
(i) Expenditure by the Intervention Board (a):					
cereals	229.4	210.1	120.6	120.8	197.3
oilseeds	208.3	177.8	187.0	236.4	250.6
sugar	151.1	118.8	104.0	110.7	111.1
beef and veal	201.4	149.5	63.8	357.6	341.0
sheepmeat	119.4	133.2	78.6	126.5	154.7
pigmeat	-1.9	0.3	2.6	6.6	3.5
milk products	251.1	48.9	92.9	203.9	165.9
processed goods	47.6	45.6	38.5	44.6	54.6
other (b)	16.6	27.2	46.7	55.2	45.4
Sub-total	1,222.9	911.4	734.7	1,262.3	1,324.1
(ii) Expenditure by the Agriculture Departments:					
Repayment of Cereals Levies	1.7	2.2	24.2	19.2	1.7
Suckler Cow Premium Scheme	36.7	37.7	61.8	73.7	70.4
Annual Premium on Ewes	80.6	130.7	114.9	201.8	276.1
Milk Non-Marketing Premiums	...	...	...	...	...
Payments to producers giving up some milk production	51.0	74.1	65.2	59.6	54.5
Beef Special Premium Scheme - Northern Ireland (c)	..	..	11.6	6.2	7.3
Orchard Grubbing Scheme	..	..	..	..	1.5
Sub-total	170.0	244.6	277.7	360.5	411.7
Total	1,392.9	1,156.0	1,012.4	1,622.8	1,735.8
II Price guarantees					
Potatoes (d)	-	-	-	-	-
Wool (d)	0.7	0.1	2.9	18.2	38.0
Assistance to the Egg Industry	...	2.9	0.1	0.1	-
Total	0.7	3.0	3.0	18.3	38.0
III Support for capital and other improvements					
Farm and Conservation Grant Scheme (EC) (e)	...	...	0.4	3.2	6.3
Agriculture Improvement Scheme (EC) (e)	18.9	25.9	24.0	13.3	10.2
Agriculture and Horticulture Development Scheme (e) (f)	32.3	21.6	15.1	12.2	7.7
Farm structures	0.4	0.4	0.3	0.3	0.3
Agriculture and Horticulture Grant Scheme (e) (g)	3.6	1.0	0.1	-	-
Agriculture Improvement Scheme (National) (e)	9.0	7.1	7.2	0.9	0.3
Northern Ireland Agricultural Development Programme (e) (h)	2.0	1.1	2.6	7.0	8.3
Farm Woodlands	..	..	0.3	1.0	2.1
Farm and Conservation Grant Scheme (National) (e)	..	..	11.5	32.1	33.8
Guidance Premiums	2.3	1.0	0.3	0.1	0.1
Farm accounts	0.7	0.4	0.2	0.2	0.3
Others (i)	0.3	0.2	...	0.2	0.4
Total	69.5	58.7	62.0	70.5	69.8

TABLE 9.1 Public expenditure under the CAP and on national grants and subsidies (continued)

£ million   April/March (financial years)

	1987/88	1988/89	1989/90	1990/91	1991/92 (forecast)
IV Support for agriculture in special areas					
Hill Livestock Compensatory Allowances:					
cattle	52.0	50.0	52.9	57.4	69.7
sheep	69.0	62.6	69.5	73.0	89.6
Additional benefit under AHDS, NIADP, AHGS, AIS (EC), AIS (Nat) FCGS (EC), FCGS (Nat)	20.9	19.6	20.1	18.6	19.8
Others (j)	4.1	4.9	8.1	10.1	11.3
Total	145.9	137.1	150.6	159.1	190.4
V Other payments					
Set-Aside	..	..	9.8	19.1	27.4
Milk outgoers scheme (h)	11.1	11.4	4.7	0.8	0.7
Weather Aid 1985	..	..	-	-	-
Storm Damage 1987	..	0.2	0.1	0.4	-
Sheep compensation scheme 1986	1.0	0.9	0.4	0.9	1.0
Cooperation grants	1.7	1.2	1.2	0.9	1.2
Crofting building grants and loans (net)	3.3	3.1	2.7	3.3	4.2
Farm Diversification:					
Capital Grants	..	0.5	2.3	2.8	3.0
Marketing and Feasibility grants	..	-	-	0.1	0.4
Nitrate Sensitive Areas	..	..	..	0.3	0.9
Environmentally Sensitive Areas	2.9	6.8	8.6	9.4	11.6
Extensification	..	..	..	-	0.4
Others (k)	1.0	0.8	0.2	1.0	1.2
Total	21.0	24.9	30.0	39.0	52.0
Total I to V (l)	1,630.0	1,379.7	1,258.0	1,909.7	2,086.0

(a) The figures are net of receipts treated as negative expenditure. Receipts from levies on the production and storage of sugar and isoglucose and on third country exports, which are regarded as Community Own Resources, are excluded.
(b) Includes eggs, poultrymeat, fruit and vegetables, hops, herbage seeds, dried fodder, peas and beans, fisheries and flax. Also includes expenditure on products covered by the CAP but not produced to any significant extent in the United Kingdom (olive oil, rice, wine, grape must and hemp).
(c) Payments in Great Britain are made by the Intervention Board and included in Beef and Veal expenditure in Section (i).
(d) Payments in respect of potatoes and wool relate partly to the crop or clip of the year indicated and partly to the crop or clip in the preceding year or years.
(e) Farmers in special areas are also eligible for additional assistance. The estimated benefit is shown separately in Section IV of the table.
(f) Includes the Farm and Horticulture Development Scheme.
(g) Includes the Farm and Horticulture Capital Grant Schemes.
(h) Except for the Northern Ireland Agricultural Development Programme (NIADP) and the Milk Outgoers Scheme expenditure from the Northern Ireland block is excluded.
(i) Includes loan guarantees, grants for agricultural drainage in Scotland and farm structure loans.
(j) Includes the integrated development programme for the Western Isles, the agricultural development programme for the Scottish Islands and grants for crofting improvements.
(k) Includes producer organisations and forage groups and Shetland wool producers.
(l) Receipts from the European Community (to which the UK contributes) are set out below (£ million). Receipts do not always relate to expenditure in the year in which they are received. Reimbursement of spending on structural measures (Section III) is normally a year in arrears. Receipts for 1988/89 and subsequent years reflect the arrangements for depreciation of stocks agreed at the European Council in February 1988.

1987/88	1988/89	1989/90	1990/91	1991/92 (forecast)
1127.7	1597.0	1228.5	1635.9	1744.6

TABLE 9.2 Public expenditure under the CAP by the Intervention Board and the Agriculture Departments - major commodities

£ million   April/March (financial years)

	1987/88	1988/89	1989/90	1990/91	1991/92 (forecast)
Cereals					
Intervention purchases/sales	-88.5	0.7	-42.3	5.9	2.8
Intervention storage costs	59.6	32.6	19.8	9.4	14.8
Export refunds	273.8	238.7	236.1	184.0	259.6
Disposal measures	26.9	25.5	18.8	22.5	27.3
Co-responsibility/additional levy	42.3	-86.8	-89.9	-82.8	-107.1
Production support	1.7	1.7	2.3	1.0	1.7
Total cereals	231.1	212.3	144.8	140.0	199.1
Oilseeds					
Intervention purchases/sales	...	-0.5	...	-	...
Intervention storage costs	...	0.2	...	...	...
Export refunds	...	...	...	0.1	...
Production support	208.3	178.1	187.0	236.2	250.6
Total oilseeds	208.3	177.8	187.0	236.4	250.6
Sugar					
Intervention storage	24.2	18.4	24.5	22.7	26.7
Export refunds	121.3	86.1	64.1	69.0	48.9
Disposal measures	5.5	6.6	5.8	4.0	7.7
Production support	-	7.7	9.6	15.0	27.8
Total sugar	151.1	118.8	104.0	110.7	111.1
Beef and veal					
Intervention purchases/sales	47.9	6.4	-4.6	266.3	231.4
Intervention storage costs	20.3	12.5	6.5	31.8	48.8
Export refunds	-2.8	16.0	19.0	20.3	25.3
Production support	172.7	152.3	116.3	119.1	113.3
Total beef and veal	238.1	187.2	137.2	437.5	418.8
Sheepmeat					
Production support	200.0	263.9	193.5	328.3	430.8
Pigmeat					
Intervention storage	0.5	0.4	-	...	0.3
Export refunds	-2.4	-0.1	2.6	6.6	3.2
Total pigmeat	-1.9	0.3	2.6	6.6	3.5
Milk products					
Intervention purchases/sales	92.8	-59.7	-12.6	62.6	0.7
Intervention storage costs	20.2	7.8	2.4	3.2	6.6
Export refunds	82.5	93.3	80.1	74.9	95.9
Disposal measures	121.2	65.4	65.7	89.3	97.3
Production support	-	-	-	-	-
Co-responsibility/Supplementary levy	-65.6	-58.0	-42.7	-26.1	-34.7
Payments to those giving up milk production	51.0	74.0	65.2	59.6	54.6
Total milk products	302.0	123.0	158.1	263.5	220.4
Processed goods					
Export refunds	47.6	45.6	38.5	44.6	54.6
Others					
Export refunds	16.1	22.3	37.3	42.5	36.6
Disposal measures	2.7	0.7	3.5	1.7	1.7
Production support	4.3	6.4	6.3	12.2	12.4
Miscellaneous	-7.5	-2.3	-0.4	-1.3	3.9
Total others	16.6	27.2	46.7	55.2	46.9
Total	1,392.9	1,156.0	1,012.4	1,622.8	1,735.8

TABLE 9.3 Commodity intervention in the United Kingdom

'000 tonnes

Commodity	1987 (a)			1988 (a)			1989/90 (b)			1990/91 (c)			1991/92 (c) (forecast)		
	Closing/opening stock (d)	Pur-chases	Sales	Closing/opening stock (d)	Pur-chases	Sales	Closing/opening stock (d)	Pur-chases	Sales	Closing/opening stock (d)	Pur-chases	Sales	Closing/opening stock (d)	Pur-chases	Sales
Wheat: feed	2,615	...	2,109	511	289	352	447	-	382	73	33	69	34	34	67
bread	-	-	-	-	14	14	-	-	-	-	1	-	1	8	9
Barley	1,179	44	619	601	335	229	707	2	160	550	131	1	680	144	223
Rye	-	-	-	-	...	...	...	-	...	-	...	-	...	-	-
Oilseeds	-	-	-	-	30	30	-	-	-	-	-	-	-	-	-
Beef: boneless	39	26	31	34	26	33	24	12	24	12	96	12	96	83	25
bone in	17	8	5	20	2	20	2	1	1	2	4	1	5	-	2
Butter	238	67	156	160	2	138	24	-	18	6	23	4	26	6	12
Skimmed milk powder	23	...	22	1	-	1	...	-	...	-	7	-	7	3	7

Closing stock (d)
1
...
601
...
...
154
3
20
3

(a) Calendar years.
(b) 1 January 1989 to 31 March 1990.
(c) 1 April to 31 March.
(d) These figures may not always equate to (closing stock = opening stock + purchases - sales) because of end of year stock adjustments arising from unfulfilled sales contracts etc., and because each figure is rounded.

Printed in the United Kingdom for HMSO
Dd293438 1/92 C20 G531 10170